BOOKS BY ROBERT S. DE ROPP

IF I FORGET THEE
DRUGS AND THE MIND
MAN AGAINST AGING
THE MASTER GAME
SEX ENERGY
THE NEW PROMETHEANS
CHURCH OF THE EARTH
ECO-TECH

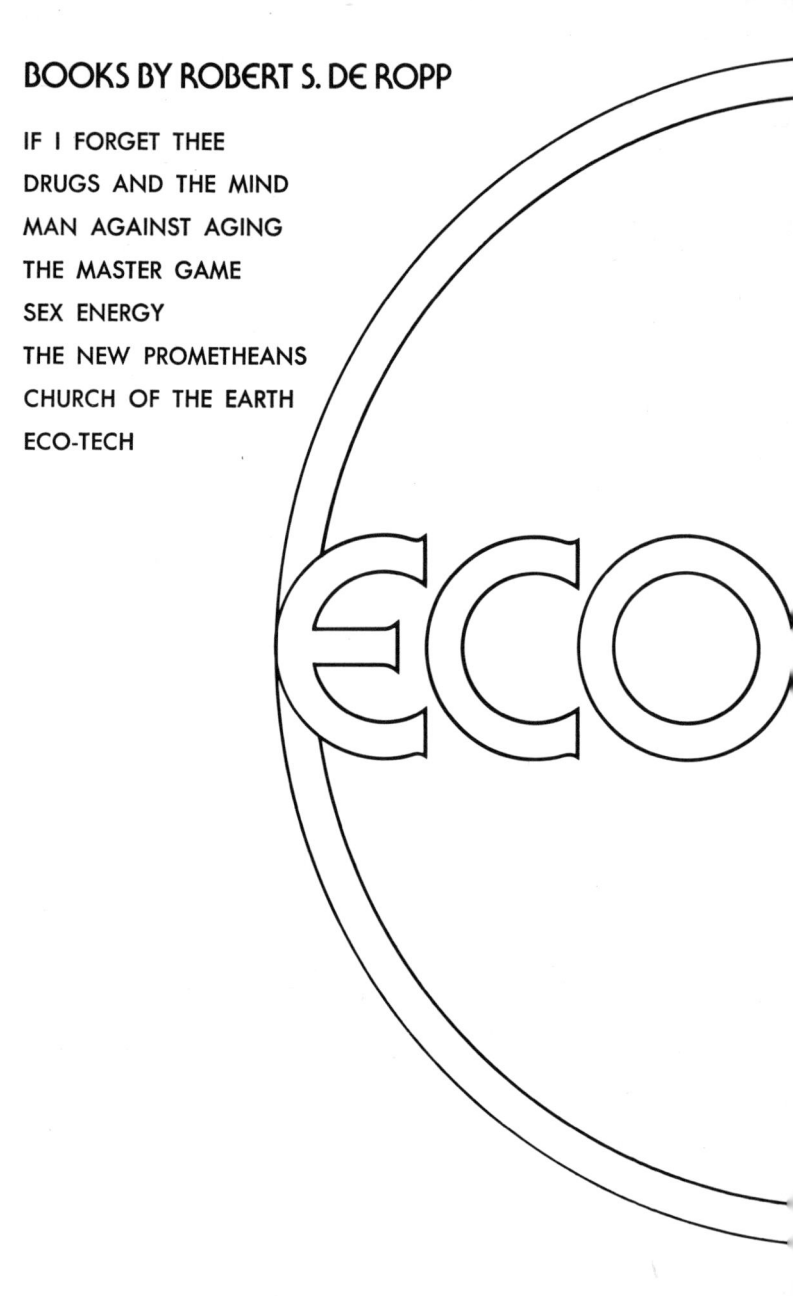

ECO-TECH

BY ROBERT S. DE ROPP

The Whole-earther's Guide to the Alternate Society

DRAWINGS BY GLENDA GUSS

A Merloyd Lawrence Book
Delacorte Press / Seymour Lawrence

Copyright © 1975 by Robert S. de Ropp

All rights reserved. No part of this book may be
reproduced in any form or by any means
without the prior written permission
of the Publisher, excepting brief quotes
used in connection with reviews
written specifically for inclusion
in a magazine or newspaper.

Manufactured in the United States of America
Designed by Jerry Tillett
First printing

Library of Congress Cataloging in Publication Data

De Ropp, Robert S
Eco-tech.

"A Merloyd Lawrence book."
Includes index.
1. Agriculture—Handbooks, manuals, etc.   2. Agriculture
—United States.   3. Home economics, Rural—Handbooks,
manuals etc.   4. Collective settlements—United States.
I. Title.   II. Title: The whole-earther's guide to the
alternate society.
S501.2.D47   640'.9173'4   75-4598

ISBN: 0-440-02233-9

CONTENTS

FOOD

FOOD GROWING 2
1. Obtaining soil 2
2. Preparing soil 7
3. Major food crops 26
4. Minor food crops 42
5. Fruits and nuts 46
6. Animal foods 51
7. Crop layout 53
8. Aquaculture 56

FOOD GATHERING 68
1. Hunting and trapping 69
2. Fishing and shore foraging 73
3. Wild plants 76
4. Water 82
5. Fire 82

FOOD STORING AND PROCESSING 84
1. Salting, smoking, drying 84
2. Canning and bottling 86
3. Pickling, jam making 89
4. Freezing 90
5. Butchering 91
6. Natural dry foods 91
7. Fermenting 93
8. Cooking 105

SHELTER II

CLOTHES *118*
1. Skins *118*
2. Cloth *122*

HOUSING *130*
1. Tents *133*
2. Tipis *133*
3. House design *138*
4. Choosing the site *142*
5. Laying foundations *143*
6. Building materials *144*
7. Log cabins *149*
8. The frame house *156*
9. Heating and cooling a house *165*
10. Water supply *168*
11. Waste disposal *173*
12. Houses on water *177*

HEALTH III

HOW THE BODY WORKS *183*
1. Digestive system *184*
2. Cardiovascular and respiratory systems *186*
3. Nervous system *188*
4. Muscular system *189*
5. Skeletal system *191*
6. Endocrine system *192*
7. Urinary system *193*
8. Reproductive system *193*

WHAT CAN GO WRONG *196*
1. Food intake *197*
2. Intake of poisons *201*
3. Infectious diseases, prevention *204*
4. Infectious diseases, immunization *207*

INTERPRETING SIGNS AND SYMPTOMS *209*
1. Gastrointestinal disturbances *209*
2. Respiratory disturbances *212*
3. Circulatory disorders *215*
4. Nervous system disorders *217*
5. Liver disorders *220*
6. Urinary system disorders *221*
7. Bone and joint disorders *223*
8. Blood disorders *224*
9. Skin disorders *226*
10. Venereal diseases *228*
11. Cancer *230*
12. Mental illness *231*

FIRST AID *233*

CHILDBIRTH *242*

DEATH *245*

ENERGY IV

SOLAR POWER *251*
1. Solar farms *251*
2. Photovoltaic cells *252*
3. Solar heaters *253*
4. Solar stills *256*

WIND POWER *259*
1. Savonius rotor *260*
2. Propeller *262*

WATER POWER *265*

METHANE POWER *270*

V CRAFTS

BASKET MAKING 277
1. Split oak baskets 277
2. Osier baskets 279
3. Coiled baskets 281

POTTERY 281
1. Working clay 283
2. Decorating 287
3. Firing 288

BEADWORK 289

WATERCRAFT 292
1. Coracles 293
2. Kayaks 295
3. Other traditional watercraft 296

BODGING AND BOWLING 297

PIT SAWING 300

BLACKSMITHING 301

INDEX 305

FIGURES

1. Compost Heap Structure 9
   Ideal Soil Structure
   Trenching
   Plowing Layout
2. Horse Plow 15
   Plow Harness
3. Pit Greenhouse 24
   Hot Frame
   Hydroponic Culture
4. Potato Storage 38
   Espalier
5. Crop Layout 54
6. Spacer for Sowing 56
7. Pond Construction 58
8. Methods of Oyster or Mussel Culture 66
9. Traps and Nets 72
10. Edible Wild Plants 77
11. Smoking and Drying Meat or Fish 85
12. Butchering Beef and Pork 92
13. Wine and Cheese Making 96
14. Solar Cookers (Maria Telkes Design) 115
15. Preparation of Wool and Flax 123
16. Spinning Wheel and Loom 127
17. Carpet Knots 130
18. Indian Tipi 135

19. House Designs *140*
    Solar House
20. Laying Foundations *144*
21. Construction Methods: *146*
    Wattle-and-Daub
    Rammed Earth
    Stone
22. Log Cabin: Foundations and Floor *151*
23. Log Cabin: Stockade and Horizontal *153*
24. Making and Laying Shingles *155*
25. Floor, Door and Window Framing *158*
26. Framing Roof *162*
    Laying Shakes
27. Framing Gambrel Roof and Dormer *164*
28. Chimney Flashing *167*
29. Water Sources and Well Drilling *171*
30. Waste Disposal Units *175*
31. Cardiovascular, Muscular and Digestive Systems *185*
32. Female and Male Reproductive Systems *194*
33. Menstrual Cycle *195*
34. Childbirth *243*
35. Solar Water Heater *254*
36. Solar Still *258*
37. Savonius Rotor Wind Machine *261*
38. Impulse Water Wheel *268*
39. Overshot Water Wheel *269*
40. Methane Generators *272*
41. Basket Making *278*
42. Slab and Cast Pottery *284*
43. Kick Wheel and Kiln *287*
44. Beadwork *291*
45. Watercraft *294*
46. Bodger's and Bowler's Lathes *298*
47. Blacksmith's Tools *302*

TABLES

Table I
Sample vegetable garden   *55*

Table II
Allowable spans for floor joists and roof rafters   *159*

Table III
Building code requirements for septic tanks   *177*

Table IV
Relations between units of mechanical work and heat   *249*

Table V
Characteristics of two solar houses   *256*

Table VI
Wind machine output in watts   *264*

Table VII
Relation of head and flow rate to power produced by an impulse-type water wheel   *266*

I FOOD

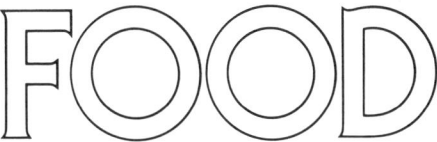

Food is man's first need. Man can live without shelter and can manage without clothes. He can certainly dispense with cars, planes, radios and television sets. But without food he dies, and so do all his fellow animals. The need to devour food links together all members of the animal kingdom and differentiates them from the green plants, which live on air, water and sunlight. In scientific jargon the animals are *heterotrophs* (organisms dependent for food on other organisms), the green plants are *autotrophs* (self-feeders).

We will discuss the nature of food and how it is used by the body in Chapter III. Here we need only point out that man can get his food in two ways. He can gather it or he can grow it. Food gathering was the way used by early man. He hunted, he fished, he collected fruits and berries. Food gathering is the method of food getting used by most animals in this biosphere. The only creatures besides man who grow their food are certain species of ants which cultivate fungus gardens.

The great Neolithic revolution, which occurred around 10,000 B.C., resulted from man's realization that seeds put in the ground would produce more seeds, and that this was a surer way of getting food than gathering it at random. He discovered he could scratch the earth's surface with a stick, bury the seed and later reap a harvest. He also realized that some animals could be trained to live with man. They would work for him, carry him, guard his property. He could kill them and eat their bodies, or eat their secretions (milk). Here again the ants were ahead. They had learned long before man did to keep herds of other species of insects (aphids in most cases). But whereas the ants were locked into certain patterns by a mode of behavior we call instinctive, man was governed by a mode we call intelligence, which is far more flexible. So methods of food growing changed. New food plants were discovered. The digging stick became the plow pulled by oxen, horses or mules. Finally the draft animals were displaced by tractors and the huge farm-factories of today displaced the individual farmer, with effects on the ecology that may yet prove disastrous.

Now we are concerned with the realities of food growing. These involve obtaining soil, preparing soil, choosing food plants, raising animals.

FOOD GROWING

1. Obtaining soil

Soil!

Every man and woman shows his/her essence by reaction to the soil. For the true Whole-earther soil is the basic reality. He loves it, revels in it, feeds it, works it, renews it. In soil he perceives the alembic of his own life, the crucible in which his substance is generated by the heat of the sun and the magic of the biosphere. The soil in his garden gives him life and the soil under his feet will one day cover his grave. He is close to the earth.

So a patch of soil is the first thing he requires. How is he to get it? Three possibilities exist. He can buy land, rent land or simply borrow it. Let us consider these possibilities.

Group caretakers. It is possible, at least here in California, for a responsible, properly organized group to take care of a ranch for someone who is too busy elsewhere, too uninterested or simply too lazy to take care of it himself. I make this statement confidently because from 1967 to 1972 our creative community occupied an 80-acre ranch on Sonoma Mountain, grew most of the food it needed on soil built up by liberal dressings of manure, repaired buildings, cared for a prolific 3-acre vineyard, two goats and assorted chickens. The owner was satisfied with a modest rent and one-third of the gross on the vineyard.

Some might object that we were exceptionally lucky to find such a place. Not so. Larry Kelly of Kelseyville in nearby Lake County wrote to *The Last Whole Earth Catalog* that he had landed a caretaker job on a 5-acre farm with free house, utilities and a salary of $125 a month and all the fruit and nuts he and his group could eat from the orchard. "I am amazed at the number of caretaker positions open in this area. Check the Santa Rosa *Press Democrat*—there are usually several caretaker ads at any given time."

So I checked the *Press Democrat*. There weren't several caretaker ads but there was one: "Caretaker for cattle ranch nr. Sonoma. State full particulars re: exp. w. cattle, horses, farm equipmt. House and good salary."

So there are such positions. They result from the fact that much if not most of California is owned by absentee landlords who have invested in land as a hedge against inflation and who have neither the wish nor the ability to use it. There are thousands upon thousands of acres around here simply lying idle, awaiting sale to some developer. Much of it is first-class agricultural land, formerly used for dairy farming. The farmers have been driven out by high taxes. The land lies idle.

> "Ill fares the land, to hastening ills a prey
> Where wealth accumulates and men decay."
> (Oliver Goldsmith, *The Deserted Village*)

The group caretaker position will work only if the group scrupulously abides by its side of the bargain. Caretaking means precisely what it says. The owner must feel sure that his property will not only be cared for but, if possible, improved. If the group has a garden it should give part of the produce to the land owner. If it works a vineyard or orchard the usual arrangement around here is two-thirds to the workers, one-third to the owner, and the workers pay the cost of cultivation, spraying, picking, etc. If you want to live as a rural commune and can't afford to buy your own land, then you must take proper care of someone else's and at least see that the owner gets enough out of it to enable him to pay his taxes. Meanwhile you get valuable experience and can find out, at a minimum expense, just who in the group is really able to pull his/her weight. It takes time for any creative community to stabilize, get rid of free-loaders, troublemakers and misfits. It is better to do this before the members sink their joint savings in a piece of property.

Land given to God. It was my old friend Lou Gottlieb (Lucky Louie Love Divine) who first startled our local bureaucrats by deeding his place, the Morningstar ranch, to God. The clerk at the courthouse who recorded the deed raised his eyebrows but did not comment. The judge, when Lou refused to pay his taxes and told the assessors to direct their requests to God, did. God, said the judge, cannot own property in California.

So much for God.

Later Bill Wheeler, owner of a much bigger ranch, followed Lou's example and opened his land to all comers. The slogan was "Land, access to which is denied none." Anyone could come and make his home there.

This situation would seem, on the face of it, to offer a perfect opportunity for a would-be creative community to acquire a land base. Here we have two idealistic individuals sufficiently free of greed for possessions to offer their land to those who have none without any talk of rental agreements, mortgages, etc. A refreshing change!

But did the new utopia bloom at Morningstar or the Wheeler ranch? Perhaps it did, for a while, but the bloom soon faded. Land, access to which is denied none, tends to attract people who are denied access everywhere else. It tends to attract outcasts. Obviously a society in which more and more people are being elbowed aside by machines and becoming technologically unemployed is going to produce outcasts in plenty. And outcasts must go somewhere. It could be argued that they should be allowed to live at Morningstar or the Wheeler ranch because it is cheaper than supporting them in the city or putting them in jail.

This issue was argued, with a good deal of passion, in court and out. The neighbors complained that the outcasts, who appeared in court wearing bells, beads and beards, did not know how to behave. They walked around naked, fucked freely, lit fires in the dry season and did not even have the sense to bury their feces. Their shacks violated the building codes, their shitting habits violated the health codes, their dogs, as untrained and uninhibited as their owners, had a way of worrying the local sheep. But the outcasts protested they were simple children of nature. They were happy to live in trees or caves or shacks or tipis. They did not want homes with stucco and swimming pools and two-car garages. They did not need the materialistic symbols which screamed, "Look at me, I'm rich."

The battle raged. Lou Gottlieb was hauled into court, thrown into jail, fined a total of $15,000. Bill Wheeler's ranch was assaulted by land and by air. It was a regular war. Later, when Wheeler tried to conform and applied for a use permit for a campground, it was denied him by the elected county officials. Nearly every spokesman at the hearing favored the campground, which would give the outcasts a chance to meet the county's legal requirements without sacrificing their way of life. The supervisors said no, 4–1.

There is a moral to this story. If you want to give your land to God and yet stay out of trouble with the local authorities, you had better appoint yourself God's deputy and make a few rules. The hippy may be one of nature's noblemen, a free spirit who has shed the fetters of possessions, a contemporary holy man, a friend of all the world. But this wandering hippy, if he wants to be part of a creative community, must at least respect a few natural laws which govern health and safety, or else various officials, appointed to protect the community at large, will surely be forced to destroy the creative community even if they sympathize with its aims.

Moreover, to qualify as a Whole-earther and a member of the

Alternate Society, the hippy must stop accepting handouts from the society he rejects. Which means he must grow his own food. Which means he must obey the rules nature imposes on all food growers: prepare land, sow seeds, control weeds, gather harvests. Otherwise the holy hippy is nothing but a free-loader and the man who shelters him is merely encouraging parasitism.

Rentals. The practice of renting a piece of land and giving a portion of the produce to the owner has been a way of life from time immemorial and a source of much injustice and misery. It does not have to be. If the owner is reasonably intelligent, he will understand that it is in his own interest to keep his tenants healthy and happy. Few landowners seem to understand this, and the careworn faces of the sharecroppers in such books as James Agee's *Let Us Now Praise Famous Men* bear witness to this fact.

Unless you know and can really trust your landlord, renting a farm has little to recommend it. What you pay in rent could be more sensibly paid to a bank as interest and principal on a mortgage. You can then really work up some enthusiasm for improving the soil, maintaining the buildings and in general adding to the value of the property.

Buying. Buying a ranch or a farm is the most obvious way for a creative community to get its land base. It's a truth, proven by many examples, that really determined people, though city-bred and ignorant of farming, can pool their resources, get land and survive.

Any would-be Whole-earther who wants to start a commune would do well to read Steve Diamond's book, *What the Trees Said,* Delacorte Press/Seymour Lawrence, New York. They did things the hard way—a group of city kids sick of trying to foment a revolution and sick of cities. So they bought a rundown farm near Montague, Mass. It had a house with 17 rooms built "not with one generation or family in mind, but with the intent to house children, grandchildren and, if the luck held, great-grandchildren." But the luck did not hold. The last of the old family died. The house and its 60-acre farm were sold. The children of the former farmers "went to the cities in search of homes heated with the flick of a switch, looking for a middle-class security that their farmer parents were never quite able to guarantee. And we, who are the children of the secure, find ourselves reclaiming that land, desirous of the poverty that helps keep us free."

One principle should be remembered by all who wish to live off their land. *Keep operations as small as possible and as intensive as possible.* Use every square foot of your garden and waste nothing. The task of soil improvement is an onerous one, involving a lot of tender loving care. It is difficult or impossible to improve a large acreage. A small plot, however, can be brought to a very high level of fertility.

But what is a "small" plot. How small is small?

A small plot is one which will yield enough to feed two Whole-earthers and their two children, this being the basic biological cell out of which the Alternate Society is constructed. It is not easy to translate this into soil area because so much depends on what the family eats. If they live mainly on vegetable food, they can get all they need from one intensively worked acre and even have a surplus to sell or trade. But if they want animal food they will need more land, unless they follow my own example and draw all their animal protein from the vast stores of this material which exist in our oceans.

If you want to grow your own animal protein you can probably manage on 5 acres. Much depends on method. With proper crop rotation and well-managed pastures it may be possible to produce the desired protein on 2 acres. Figuring minimum areas for production of a one-year food unit (the amount of food eaten by the biological unit in a year) is one of the Eco-Tech research projects. Now we say only "Think small." If you want privacy, buy 50 acres or 100. But don't try to farm it all. You will merely damage the soil, work yourself to death and pollute the atmosphere with tractor exhaust. Take from the earth only what you need, give back to the earth what you have done with.

2. Preparing soil

Once the group has its soil, whether it has bought it, rented it or borrowed it, it can begin the work of soil improvement. There are a few soils that need no improvement. The rich black soils called *chernozems* that occur in some parts of the world are so naturally fertile that they need no additional fertilizers. However, even such soils must be treated with love and care to prevent erosion and destruction of soil texture. Other soils, poor to begin with, must be substantially reconstructed in order to bring them to a high level of fertility. To do this a food grower must understand what it is plants take from a soil and how they take it.

Topsoil, in which plants will grow, is the result of centuries of work by such agents as rain, wind, ice, snow, lichens and the plants themselves. Consider a bare rock. The only things that will grow on it are lichens (which are actually two plants, a fungus and an alga) and perhaps moss. Little by little the rock cracks as a result of frost or rain. In the cracks the remains of lichens and mosses accumulate. Seeds fall into this debris, germinate, send out roots. The roots crack the rock some more. The cracks become deeper and more numerous as more plants settle on the rock. Finally the rock falls apart, disintegrates. Its fragments, mixed with plant remains, form soil (fig. 1). The process, starting with the rock, takes millions of years.

So soil is a mixture of three things: very fine particles (clays), coarser particles (sand or rocks) and plant remains (humus). Humus is the organic fraction of soil and it forms a link with the clay particles. The center of the fertility of a soil is the clay-humus complex to which nutrient minerals become attached. This is the life of the soil (fig. 1).

A soil may be lifeless, incapable of supporting plant growth, for several reasons: (1) Too much sand. A sand dune is a good example. The only things that will grow on it are a few tough plants called *xerophytes* that can stand drought and hold on to the shifting base. (2) Too much clay. Very heavy clay is a dense substance. In wet weather it absorbs water and becomes airless and soggy. In dry weather it sets like cement. (3) Too much humus. A peat bog is an example. You can't grow a garden on a peat bog. The soil is too wet, too acid and too low in oxygen.

The surest way to gauge the fertility of a piece of soil is to look at the weeds. A rich, heavy weed growth indicates a fertile soil. The plants to look for are nettles and thistles. Both plants like a lot of nitrogen and where they grow well the soil is naturally fertile. Patches of swamp grass indicate bad drainage. Sorrel, with its reddish foliage, indicates an acid soil. The poorer the weed growth, the more infertile the soil.

Now for some remedies. Take a careful look at the soil texture. Turn the soil over, pick up a handful. How does it feel? This is getting down to earth, the hallmark of Whole-earthers every-

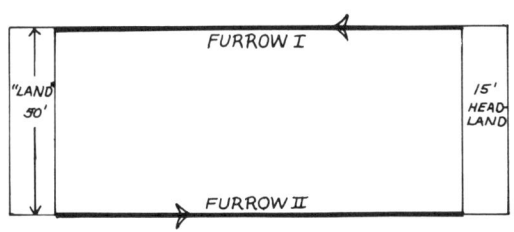

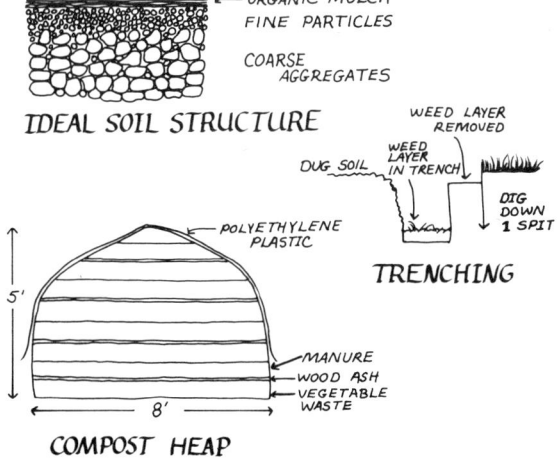

FIGURE 1

where. One must learn to talk to the soil, smell it, feel it, squeeze it in one's hands. The overlight soil falls apart, the overheavy soil squeezes into a soggy mass, the perfect soil holds together but remains friable. It smells just right, the characteristic smell of humus.

A thin soil has too much sand and not enough humus. It tends to lose its nutrients by leaching. The rain washes them out. The quickest way to fatten a thin soil is to apply loads of manure. This treatment will also open up a heavy clay. The problem here is where to get the manure. We get it from our few surviving dairy farmers, rapidly being pushed out by the developers. We give them vegetables in exchange.

If you keep your own livestock, goats, cows, chickens, rabbits, all their shit should go back into the soil. It is advisable to compost it first. This certainly should be done if you use human shit. Use of the latter, if the local health department hears about it, will generate a lot of static. This is understandable. Human feces can carry some mighty unpleasant organisms, including those of typhoid fever, bacterial and amebic dysentery, infectious hepatitis, cholera and a variety of parasitic worms. It really is not good stuff to spread around unless it has undergone sterilization.

One of our Eco-Tech projects involves working out methods of processing human wastes so that they can be returned to the soil without loading it with pathogens. Anyone who has lived in countries where human wastes are used to fertilize the soil knows what a menace they become. To eat raw fruits or vegetables in Mexico without first soaking them in Clorox is to ask for a bout of dysentery. Those who would use human wastes should therefore take precautions. For more on this subject see Chapter III.

Composting. Composting is a method of reducing organic matter to a form in which it can readily be incorporated into the soil. If you want to do the job properly, build a composting shed and place it as near as possible both to the barn and to the garden. The shed is to keep off rain. Rain soaks the compost, leaches out essential nutrients.

Build your compost heap from alternate layers of vegetable

waste and manure (fig. 1). Throw in any wood ash and bones. Make the base no more than 6 × 6 feet and 5 feet high. If the material is dry it must be wetted as the heap is built, but the amount of water must be right. Too much and you get a soggy mess, too little and you get no heating.

Heating is the sign of a living heap. The heat is due to activity of billions of bacteria who have the capacity to withstand temperatures that would kill most organisms. The combination of high temperatures and bacterial action breaks down the plant material to humus. The pile becomes black, compact and sweet-smelling. The process, however, takes place only within the heap. Stuff on the outside remains unchanged. Every heap therefore must be turned, the material on the outside put on the inside. If in the process it has become too dry it can be rewatered.

Surface mulch. Composting involves a good deal of hard work and is possible only if animal wastes are available. Surface mulching with plant waste may be an equally good or even better way of building a soil. This is the contention of Edward H. Faulkner, who has used the method and described his results (*Plowman's Folly: A Second Look, Soil Development,* University of Oklahoma Press, 1963).

Faulkner was a real Whole-earther long before this term was invented. He grew up in an area where the small subsistence farm still exists, the Appalachian Mountains in Kentucky-Tennessee. His father grew fine vegetables with the help of manure and fertilizers. The son went one better and grew them without either.

Ed Faulkner certainly made a big hit with *Plowman's Folly* and anyone interested in this development should read it. What he claims, in brief, is that the moldboard plow, by turning vegetable wastes under the soil, puts them in a place where they do the least good. The best place for vegetable wastes is on the surface. So, he suggests, stop plowing. Leave the corn stalks, stems, leaves, wastes of all kind on the surface and chop them up to form a mulch. They give a nice workable surface layer and help preserve soil moisture.

There is much to be said for the practice. The idea behind it is that you grow your soil while growing your crops. Soil is a living organism and it needs feeding. For this reason it is necessary, when soil is poor in humus and compost is not available, to grow crops to feed the soil and then to work them into its surface. This is one function of cover crops.

Here in northern California cover crops are essential. The winter rains can be torrential and anyone foolish enough to leave his soil unprotected will see much of it wash away. Winter temperatures, however, are mild and suitable cover crops grow throughout the winter. A crop of oats or mustard will be knee-high by the end of February.

This lush growth can be eaten down by goats or rabbits in March or it can be disced into the soil. Or it can be cut with a mower and left on the surface, or worked over with a Roto Tiller. But it should not be plowed under. It does most good on the surface. If it obstructs seed sowing, rake it to one side but pull it back lightly over the seeds after they have been sown.

Long-term soil development may require more than winter cover crops which, in any case, will not grow in areas with severe winters. One of the finest soil builders is sweet clover, a tough, deep-rooted plant that survives with very little water. It has two great virtues. Its deep roots bring nutrients to the surface from the lower soil levels. Its roots fix nitrogen. Sweet clover is a biennial, grows a leafy base the first year, flowers in the second. You can pasture cows, goats, rabbits on it in year two. *But* it has one drawback. It is tough and therefore hard to control. The root of a well-grown, two-year-old plant resists anything less rugged than a mattock. It will seed itself everywhere and go on producing new plants for years. These must be hoed out at the seedling stage or they become a problem. Oats, rye, mustard and other lush-growing plants can also be grown as soil foods, but do not fix nitrogen.

Working the soil. In the United States vast amounts of money and energy are used year after year to push millions of tons of soil to one side and then push it back again. Just how much energy this involves can be estimated by any Whole-earther willing to dig his own vegetable plot. He will have to work hard to

FOOD GROWING   13

turn over a plot 30 × 30 feet to the depth of a foot in a day. Is all this labor really necessary?

The answer is, this depends on the soil. Plant roots must have oxygen and they get the oxygen from the pores in the soil. An ideal soil is spongy and well aerated. About one-third of its pore space should be occupied by air and two-thirds by water. If it is completely waterlogged most crop plants will fail. Of all crop plants grown, only a few (rice, watercress, water chestnut, etc.) will survive on water-saturated soil.

Winter weather with heavy rains, snow and slush leaves the gardener facing a new growing season with a soil that may be completely waterlogged. If it remains waterlogged he must take steps to loosen it. The old-fashioned gardener, a species hardly known in the United States, will trench his soil, opening the trench to the depth of a spit (about a foot), burying surface weeds, then forking the soil over them (fig. 1). This is glorious work, a deep and satisfying rhythm, building strong muscles, satisfying the old hungers. Dig. Dig it, man! And if you get an aching back, reflect that this has been man's fate since Adam and Eve were thrown out of the Garden of Eden.

Digging will certainly fit your soil and leave it well aerated and leave you well aerated too if you breathe properly. You can, if you have too much soil to dig and don't mind befouling the air with gasoline breakdown products and noise, use a small tractor or Roto Tiller to produce much the same effect. Or if you follow Faulkner and shun moldboard plows you can use discs. The ideal soil structure is one in which rather large lumps are overlain by smaller fragments which in turn are overlain by an organic mulch. This is the perfect seed bed and for fine seed like that of carrot or onion a perfect seed bed is the key to a good crop.

Can one fit soil for a crop without digging or using a Roto Tiller? This depends on the soil and the arrangement of the plots. Any Whole-earther who wishes to experiment can try the method of raised beds. This involves preparing a series of raised beds not more than 5 feet wide preferably on a south slope following the contour of the land. The secret is that these beds are never trodden on. The gardener reaches over from the side. If he has to put his foot on the bed he uses a board to distribute his weight. He protects the bed always with mulch in winter, hoes, rakes but does not dig. The method is fine for very intensive gardening and is effective if one has a good loam to begin with. A heavy clay makes this method impractical until great amounts of humus have been incorporated into the soil.

Soil-working tools. These tools, the lineal descendants of the Neolithic digging stick, are among man's oldest and most sacred instruments. They should be treated with respect, cleaned after use, not left out in sun or rain. The Whole-earther whose garden plot is small and who loves nothing better than a good spell of digging can manage with a digging fork, a spade, a hoe and a rake. He will also need a trowel for planting, a line (builder's twine), a measuring rod and a scythe or sickle for cutting down weeds. If he needs a small tractor the Gravely is much recommended (Gravely Corp., Gravely Lane, Clemons, N.C. 27012) or the Troy-Bilt Roto Tiller (Troy-Bilt, 102nd St. and 9th Ave., Troy, N.Y. 12182).

Or he can buy a horse or a mule and enjoy himself by using a beautiful streamlined moldboard plow (Ed Faulkner notwithstanding). It may not be the best possible thing for the soil, but I can say from experience that the sensation of two powerful horses pulling the plow through the earth and the feel of that power transmitted through the plow handles is one of life's pleasures. An acquired taste perhaps, and not every man's cup of tea, but definitely something that every Whole-earther should know about.

A moldboard plow must be used intelligently (fig. 2). It is an instrument for turning soil over and will not do its work if wrongly set. The plow's essential parts are the colter, the plowshare, the moldboard and one or two wheels. The colter is usually in the form of a disc. It cuts vertically into the soil to almost the depth of the furrow. The plowshare is a blade of hardened steel that cuts horizontally and determines the width of the furrow. The moldboard turns the soil over. The wheels determine the depth to which the plowshare penetrates.

FOOD GROWING   15

A moldboard plow can turn soil either to the left or the right. Most plows are set to turn right. Before you start plowing, lay out *lands*. These are strips of land to be plowed in one piece, either sides to middle or middle to sides. If you start in the middle, mark off the center point at each end of the land, plow a straight furrow from one to the other, come back on the other side. At the end of the land you must leave a *headland* wide enough to turn your plow (about 15 feet). The headland can be plowed later when the piece is completed (fig. 1).

If you plow middle to sides you end up with a ridge in the center and two furrows at the edges of the land. If you plow

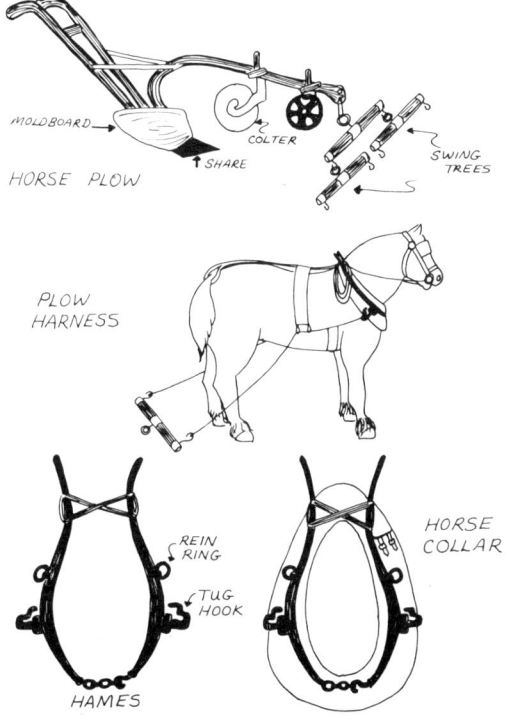

FIGURE 2

sides to middle you end with a furrow in the middle of the land. Alternate the procedure from year to year.

In contour plowing you leave a strip unplowed between each land to prevent soil erosion. You can plow two lands at a time, turning the soil uphill on one, downhill on the other. Always alternate plowing direction from year to year.

A horse plow must be steered to keep the furrow straight. Pressure on the right handle will cause the plow to swing to the left. Strong pressure on the right handle will cause the plow to come out of the ground. This maneuver must be practiced so that the plow is out of the ground at the headland. The plow can be dragged on its side until the horse or horses are in line with the next furrow, then straightened out and let into the soil at the right moment.

The harness for a two-horse plow is shown in fig. 2. Two swing trees are connected to a main swing tree, the plow chains are hooked up to the *hames,* which in turn are attached to the horse collar. If you use two horses one will walk in the furrow, the other alongside the first. On heavy soil two horses will be necessary.

Remember this. If your land slopes, plow on the contour. And if you have steep slopes, always leave strips unplowed to hold the soil. Soil erosion is the curse of American agriculture. Millions of tons of topsoil have been scoured from the land and ended on the ocean bottom because a few simple principles were neglected. Get advice from the Soil Conservation Department. If you want to plant a garden on a steep slope (because it's all you have or you like slopes), face the fact that you may have to terrace it. There is land in Java so steep that, with American agricultural methods, it would have eroded away in one year. Yet the Javanese have farmed it for centuries. How? By terracing. Every tiny plot is handled as a precious possession. It's life or death for them. They don't have a whole continent to ravage at will. The day will come when we won't have one either. So plow on the contour, leave strips of sod and, if need be, terrace.

Horses. Draft horses, as opposed to riding horses, are massive animals which may weigh a ton or more and stand up to 6 feet at the shoulders. In his book *The Gentle Giants* (A. S. Barnes and Co., Cranbury, New Jersey, 1971) Stanley M. Jepson assures us that these splendid beasts are once more increasing in numbers in the United States. True Whole-earthers like Harry Bresley of Ord, Neb., have stuck with horses right through the great era of mechanization. Mr. Bresley farms 2,000 acres and is convinced that a young farmer, if he wanted to, could take a quarter section and make a good living off it if he would return to horses and horse-drawn equipment. As the Fossil Fuel Failure draws nearer and the price of gasoline rises there is more and more reason to agree with Mr. Bresley's opinion. Horses can live off the land; they do not need imported fuel. Furthermore, they are, on the whole, friendly beasts with which a man can establish a real emotional relationship. Who could establish such a relationship with a tractor, noisy and smelly and polluting the atmosphere with its exhaust?

Draft horses belong to five main breeds: Belgians, Clydesdales, Percherons, Shires and Suffolks. Belgians and Percherons are the most popular breeds in the United States. Paul L. Murphy of Bloomington, Ill., who breeds Belgian draft horses on his 312-acre farm, states that the Belgian is now the favorite breed. Certainly these horses have much in their favor. They are extremely strong, "blocky" and heavily muscled. The stallions can become veritable giants (Jepson publishes a photograph of one, Brooklyn Supreme, that weighed 3,200 pounds and was thought to be the biggest horse in the world). Only the Shire equals the Belgian for sheer size, but Shires, like Clydesdales, have the disadvantage of a wealth of hair, or "feathers," on the lower legs. This hair tends to become caked in mud during the winter and is a nuisance to clean.

The Percheron is not such a massive horse as the Belgian or Shire. Weights of the stallions range from 1,900 to 2,100 pounds. Percherons are commonly considered the most handsome of all the draft horses, having sleek graceful lines, neither too blocky nor too rangy. Many people think of them as being typically dappled gray but actually they are more usually black.

The Clydesdale, which ranks third in number among the heavy types of draft horses, is a somewhat lighter horse than the Percheron or Belgian. The average stallion weighs from 1,700 to 1,900 pounds and the mature mare weighs 1,600 to 1,800 pounds. These horses are beautiful beasts and have the added advantage of being, in most cases, extremely placid and gentle. Clydesdales lack the extreme blockiness of the Belgians and they share with the Shire the "feathering" on fetlocks and legs. They are typically bay or brown with white faces and considerable white on feet and legs. The Anheuser-Busch Co. of St. Louis used Clydesdales to pull its brewery wagons and has done much to popularize the breed. Its specially bred Budweiser Clydesdales average 2,000 pounds and stand 72 inches at the shoulder.

The other two breeds, Shires and Suffolks, or Suffolk Punch as it is sometimes called, are less popular in the United States than the three breeds already described. Shires are very large animals, the stallions weighing up to 2,200 pounds. They resemble Clydesdales but are consistently more massive.

Suffolks are clean-legged, compact and extremely hardy. They are the smallest of the draft horses, weighing between 1,600 and 1,800 pounds. They are always chestnut in color, short-legged with massive bodies. They are long-lived and tough, able to survive on rations that would not suffice for the heavier breeds of draft horse. Suffolks are not widely used in the United States. The breed deserves to be better known.

The average Whole-earther trying to start his own farm may not wish to buy pedigree horses, though the comment of Harry Bresley is worth bearing in mind: "I decided so long as I was going into the horse business I should go into the purebred business because I figured it didn't cost much more to raise a purebred than just a nag." The fact remains, however, that one has to pay a lot for a pedigree and that anyone with a feel for horses can often find a useful animal of mixed parentage which will work just as well as a purebred and cost much less.

How does one select a horse? The question is a difficult one to answer because so much depends, in the selection of livestock, on one's having a "feel" for animals. In general, the qualities to look for in a draft horse are short stocky legs, well-developed muscles, large girth and well-placed feet. For general farm work the horse should be from 5 to 10 years old, should weigh between 1,300 and 1,800 pounds. The temperament of the animal is every bit as important as its physical constitution. Excitable, nervous, ill-tempered horses are always troublesome to work with and can be downright dangerous.

No horse can work well in ill-fitting harness. By far the most important item of the harness is the horse collar, which should, if possible, be made specially to fit the horse. Pressure is exerted by the animal's shoulders against the pad of the collar, but the collar must not be too tight around the neck. There should be room at the base of the collar to insert the palm of the hand and enough space at the side of the neck to insert two fingers. The hames should be so adjusted that the hooks to which the traces are attached are about three-quarters of the distance from the top to the bottom of the collar. The bridle must be adjusted so that the bit does not wrinkle the horse's mouth and the chin strap should not be too tight. Belly bands must also be adjusted so that the flat of the hand can be inserted between the band and the belly. Harness should be kept clean and dressed with neat's-foot oil. Harness sores will develop if any part of the harness is wrongly adjusted. As soon as they are noticed steps should be taken to prevent further chafing. They most frequently develop on the shoulders from a badly fitting collar but may appear around the mouth, on the back or under the belly. In each case chafing is the cause.

Horses should be housed in clean, well-ventilated stalls with access, if possible, to a yard. In mild climates they can be let out to pasture all year. For a working horse of about 1,500 pounds a diet of about 24 pounds of hay and 18 pounds of oats per day is necessary. Oats are given at lunch and hay is left in the stall at night. A full hour should be allowed for lunch and the horse given its food in a nose bag. In very cold weather the horse should be blanketed during the lunch hour. It should not be watered in hot weather until it has had a chance to cool off. The grain ration should be cut down on weekends, when the horse is not working. Substitution of bran for oats on weekends will help keep the horse in good condition.

Mules. The mule is a hybrid, having a jackass or jack for sire and a mare for dam, a beast "without pride of ancestry or hope of posterity." In spite of this, many an old-style farmer, especially in the warmer regions of the United States, will prefer a mule to a horse. Its admirers declare that, in comparison with the horse, a mule will live longer, endure more work and hardship, require less attention and less feed, be less liable to digestive disorders, lameness and disease, and be more capable of performing work in the hands of a mediocre master. The mule's enemies will hasten to describe the other side of the picture and to point out that mules are stubborn, suspicious and liable to take time out for a roll in quite inconvenient places. For the true Whole-earther, eager to be free of the air-polluting gasoline-guzzling tractor, the mule may seem a good choice because of its hardiness and ability to live on poor feed. Also, if proper care is taken of its feet, a mule need not be shod.

Mules, like horses, come in various sizes. The heaviest draft mules weigh as much as 1,600 pounds. There are sugar mules, cotton mules, tobacco mules which range in weight from 750 to 1,300 pounds. The weight of the animal determines how much it can pull. A good weight for a farm mule is about 1,150 pounds.

Mules reach maturity at 5 years but can be given light work when 3 years old. A mule is at its prime between 5 and 8 years. When buying a mule the Whole-earther should look particularly at the hindquarters, as this is the business end of the animal so far as work is concerned. In a poorly formed mule the croup may be short and steep, the hips too sloping, the thighs too narrow, the bone too light and the hind legs sickle-shaped. These qualities indicate a weak animal. Conformation, however, is not everything. The mule buyer must also form an estimate of the animal's temperament. The natural tendency of a mule is to be lazy and obstinate and if these qualities are developed to excess they may make the animal worthless, however good its physical conformation.

The principles used in the correct feeding of horses may be applied to the feeding of mules. Generally, 1.1 pounds of grain and 1.25 pounds of good hay per 100 pounds live weight per day will be the right amount of feed for an animal at medium work. For heavy work, increase the amount of grain to 1.25 or 1.3 pounds.

When mules appear excessively warm after work they should be allowed time to cool before being watered or fed. The hay ration should be fed mainly at night to give the mule time to digest this bulky feed before going to work. In summer mules should be turned out to pasture after they have cleaned up their feed. In winter idle mules can be kept in good condition largely on hay and straw and a small amount of grain. Idle mules thrive better if kept in the open, preferably on pasture.

Fertilizers. Many Whole-earthers have acquired the opinion that there is something inherently wicked about using chemical fertilizers, just as the squares have acquired the opinion that there is something inherently wicked about smoking marihuana. But opinion, as the Sufi poet says, is a bird with one wing. We need fewer opinions and more knowledge.

The facts about chemical fertilizers are very simple. They are substances taken from the earth or fixed from the air. They contain three elements that no plant can do without: nitrogen, phosphorus and potassium (potash). These three are symbolized by the letters N, P and K and their proportions in a mixed bag of fertilizer are given by figures in that order, as 10-10-10 or 5-10-5.

It is a fact proven beyond question that no plant can grow without NPK and that, if any one of these elements is in short supply, growth will be limited. There are other elements plants need, including calcium, sulfur, iron, magnesium, boron, copper, zinc, manganese and molybdenum. Most soils contain enough of these last nine elements but the three major elements, NPK, are often deficient. Nitrogen deficiency can often be remedied by growing legumes on the soil. These plants, including peas, beans, clovers and vetch, fix nitrogen from the air in their roots. But legumes cannot provide phosphorus or potassium. They often need more of these elements themselves as well as a dressing of lime (calcium carbonate) if they are to grow well.

How can one tell if plants need more potassium or phosphorus? The surest way to find out is to experiment. Divide your land into small plots, give K to one, KP to another, NKP to a third, nothing to a fourth. Replicate the set several times and scatter the plots at random. Grow some average crop, say wheat, on all the plots, harvest it, weigh the crop. See which treatment gave the best result.

You will then have facts to go by and can use fertilizer intelligently.

Can manure replace fertilizer? Of course it can. At Rothamsted Experiment Station they have the oldest experiment in the world; a field (Broadbalk) that has been under wheat continuously for over 100 years. It is divided into three plots. Plot I has no fertilizer of any kind, plot II gets manure, plot III fertilizer. What are the yields after 100 years? Plot I on totally unfertilized soil gives about 12 bushels of wheat per acre, plot II with manure gives about 32 bushels, plot III with fertilizer 44 bushels.

Those Whole-earthers who insist on "organic gardening" will still get good yields if they can get good manure. But if they cannot, they had better dip into a bag at least for P and K, and for N they should plant a good crop of beans (which may need P and K but no N; a dressing of 0-10-5 can be tried).

Nothing counts in the crop growing business except results and he who would dispense with chemical fertilizer and whose soil needs P and K must get them from somewhere. Bonemeal will supply phosphorus, and wood ash will supply potassium (this is the derivation of the name potash). But bonemeal is an expensive way of getting phosphate and wood ash is in short supply. So superphosphate from rocks and potash from mines may be the only source.

Needless to say, Eco-Tech deplores loudly and emphatically the insane wastefulness of city sewage systems that pour precious phosphates and potash from the wastes of the collective man-gut into rivers, lakes and ultimately the ocean. The world's potassium and phosphorus resources are not unlimited and we can ill afford to waste them. Methods of recovering these materials and returning them to the soil *without* a host of human parasites are among the most urgent of Eco-Tech research projects. Even a corpse should not be allowed to lock up its phosphorus in a coffin. Put your elements from your wastes back into circulation. You no longer need them and they didn't belong to you in the first place.

Greenhouse. A garden without a greenhouse is only half a garden. So many advantages are conferred by a small amount of heated growing space that any Whole-earther will find it worthwhile to create such space. It need not be expensive. If you have a steep south slope on your property, simply dig out a growing pit (fig. 3). This is nothing more than an excavation in the side of the slope, with its sides and back reinforced with bricks or cement. Over the opening can be placed several large windows, which can often be obtained cheap from a wrecking yard. If you want a cheap source of heat in this miniature hothouse, spread a thick layer of fresh manure under the soil in which the plants are to grow. And on very cold nights, put a layer of insulation over the windows. On clear sunny days, be sure to raise one of the windows. Too much heat is bad for the plants. The temperature in the hotbed should not rise above 85° F (fig. 3).

The pit greenhouse is a further development of the hot frame. It is also best built on a south slope. It offers a minimum surface to the cooling winds, yet traps plenty of light. In place of glass, vinyl plastic can be used.

The greenhouse can be used for raising plants of species too frost-sensitive to be planted out directly in the soil. Examples are tomatoes, peppers, eggplants and melons. It may even be advantageous, in areas with very short growing seasons, to start winter squash and cucumbers in the greenhouse. These sensitive plants can be further protected after planting out, by covering them with "Hotkaps," which are dome-shaped pieces of translucent paper that act like miniature greenhouses.

24 FOOD

Seedlings can be grown in shallow wooden boxes (flats) or in individual containers. Seeds should be sown in early March, then transplanted to prevent overcrowding. Potting soil can be made by mixing equal parts of garden soil, sand and peat moss, which should be sifted through a ¼-inch screen. Care must be taken to avoid overwatering. Soil should be moist but not waterlogged.

The greenhouse can also be used to provide a continuous supply of greens such as lettuce during the winter months. Growth will be hastened if soil is warmed by a buried electric heating cable specially prepared for this purpose.

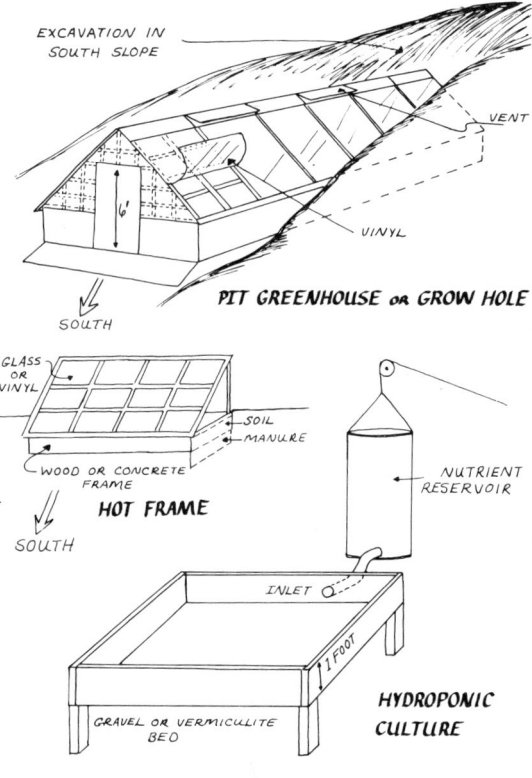

FIGURE 3

Soilless culture. Can one dispense with soil entirely? The answer is yes, but a true Whole-earther would hesitate to do so. One can do it only if there is at one's disposal the resources of a large chemical industry. Then one can, from a mixture of salts, create a nutrient solution that contains everything a plant needs. The solutions are rather complex (for details, consult *Hydroponic Culture of Vegetable Crops,* University of Illinois, College of Agriculture, Urbana, Ill. 61801). Various firms sell ready-made mixes (see *Last Whole Earth Catalog,* p. 59, for list). This technique of growing plants in solutions is called hydroponics.

Actually it is not advisable to grow the plants in the solution itself. A better method is to grow them in sand. Wet the sand with the solution and allow it to drain back into the container. This can be easily accomplished by having a reservoir that can be moved up and down or incorporating a pump into the system (fig. 3).

The technique can be used in desert areas by the sea in conjunction with a plastic greenhouse and a method for distilling sea water. It has been used at Puerto Penasco on the Gulf of California. The heart of the system, however, is a 60-kilowatt Caterpillar diesel generator, a monster that many Whole-earthers might not wish to have around. The Puerto Penasco unit certainly uses the monster very efficiently. The heat it generates distills 6,000 gallons of fresh water per day, part of which is used for hydroponics, part for drinking. The exhaust gases, after purification, are pumped into the greenhouse to raise the carbon dioxide level of the air. The plastic greenhouse, costing 15 cents a square foot, cuts down water loss by evaporation. As there is unlimited sunlight the plant growth is prodigious.

There are 18,000 miles of shore line on this planet that are uninhabitable because fresh water is lacking. So, if the diesel generator could be replaced by a solar still, this brand of hydroponics could be made the base for numerous ocean-side communities. This is a research project worthy of Eco-Tech and any creative community with a taste for fishing, sunlight and sea air should give it careful consideration.

3. Major food crops

We call those food crops major which can provide all the building blocks and the fuel which the man-body needs. This is a noble group of plants and worthy of worship as the intermediates between man and the sun, the giver of life. Without them we could not exist. The city-bred man-swarm is, of course, unaware of their presence, as it is of almost everything, but the Whole-earther, to be worthy of the name, must be acutely conscious of these marvelous forms. Their cells perform feats of photochemistry of which our own bodies are quite incapable. Within them take place the energy transformations on which the entire animal biomass of this planet depends for its existence. There is no organ more precious than a green leaf. Leaves should therefore be handled with respect and not wasted. Every Whole-earther should learn to pass the lettuce leaf test described in *Four Changes:*

A monk and an old master were once walking in the mountains. They noticed a little hut upstream. The monk said, "A wise hermit must live there." The master said, "That's no wise hermit, you see that lettuce leaf floating down the stream, he's a Waster." Just then an old man came running down the hill with his beard flying and caught the floating lettuce leaf.

The waste of food in this country verges on criminal insanity. The food thrown out daily from stores and restaurants of one major city would suffice to feed an army. Overeating is the rule. A man whose main activity during the day is to heft his own weight from one office to another takes in as much fuel as would a logger working in the woods. We gain nothing from such hoggishness except paunches, pouches and blocked arteries. Let the true Whole-earther eat sparingly, remembering the fact (scientifically proven) that the thin rats bury the fat rats, remembering also to respect the sacred plants that provide his food.

Legumes. Every Whole-earther should love the legumes. They are real benefactors. They have developed one of the most extraordinary tricks of all organisms in the biosphere, teaming up with certain bacteria which they house in special nodules in their roots. These organisms perform a feat which any chemist with a shred of imagination must regard as miraculous. They oxidize nitrogen from the air. If a chemist wants to oxidize nitrogen he uses fierce temperatures and huge pressures. The humble bacteria do it with nothing more than a few enzymes!

Legumes have flowers of exquisite symmetry. The sweet pea, which combines a beautiful blossom with one of the most delicate fragrances in the plant kingdom, shows us the general pattern, a boat-shaped keel of petals housing the stamens and ovary, two winglike petals on either side and a larger standard petal above. The purple, heavy-scented flowers of laburnum, the hanging cymes of white or yellow acacia, the flowers of beans, vetch, clover all follow this pattern. They are called Papilionaceae from the resemblance of their flowers to the butterfly. A different group of legumes includes the mimosa with its feathery, sensitive leaves that close at a touch and its masses of yellow blossoms that glorify California in February.

Green peas and fava beans are the first legumes a Whole-earther should sow in his garden. Here in California you can sow them in November if your soil is well drained. Both peas and fava beans will stand light frost. In colder climates they should go in in March as soon as the soil is workable. They both hate heat, so early sowing is essential.

We grow as much as 24 hundred-foot rows of green peas. The rows are spaced alternately 2 and 3 feet apart, the seed sown thickly in wide seed drills at the rate of 1½ pounds per hundred-foot row. We grow Little Marvel only, a dwarf pea needing no support. Sown in November it will be flowering by March. Additional sowings in January, February and March should be made or the Whole-earther will be buried in peas in May–June. As pea seed is quite expensive and a lot is needed we leave two rows unharvested, which gives enough seed for next year's crop.

The fava beans can be sown 6 inches apart, 2 feet between rows. I have used them in combination with green peas, sowing beans in November between the rows of peas, sown in February. The beans give the peas protection and something to climb on.

You must leave every third row empty or you can't get in to pick the crop.

After picking and shelling your peas (shelling takes twice as long as picking), dip them one minute in boiling water, cool and freeze. Be careful to pick them at the right stage before the sugar turns to starch. Pods should be firm and well packed. Pick as early in the morning as possible.

The fava bean seems less popular in the United States than in England, where it is called the broad bean and much esteemed as an early vegetable. The trick is to pick the beans young when they are still tender. Allowed to ripen they become very tough. They can then be ground and used to thicken stews.

Beans come in a vast variety of shapes, sizes, colors, flavors. One can make colorful necklaces of beans, which can always be boiled and eaten as a last resort. Times of sowing vary with climate. You need to wait until danger of frost is over. The pencil wax bean, Italian bean, Kentucky Wonder and Scarlet Runner are all picked while the pods are fleshy and the seeds immature. Kidney beans and white navy beans are allowed to ripen. They can be husked on a good warm day when humidity is low by stomping on the dry pods and separating the beans from the husks in a good wind. These beans can be baked with tomato sauce and are strength givers and body builders.

The soybean is a noble plant but needs good hot summers to ripen properly. So does the peanut, that odd legume that has developed the trick of burying its seed pod underground. Both of these legumes are rich in protein and also contain much oil. They can be sown in May, 2 feet between rows, but here in northern California, 50 miles from the ocean, they do not get enough heat to ripen. So we have to buy our peanut butter, which we consume in vast amounts. Great regrets!

The lima bean does better. In fact, this magnificent bean is my main source of vegetable protein. I grow the climbing variety on 7-foot poles, two 50-foot rows planted 4 feet apart, the poles sloping in and tied together. Part of the crop is harvested green and then frozen, the rest is allowed to ripen. The dry beans are delicious cooked in tomato sauce with a liberal seasoning of chili powder. These and the fish I gather from the ocean give my body all the protein it needs.

The lima bean is blessedly free from disease but several crops of kidney beans here have been rendered almost worthless by bean mozaic. This virus disease is carried in the seed and shows early in the plant's development as a yellow mottling of the leaves, which become distorted. All plants showing this symptom should at once be pulled up. Best way to get virus-free seed is to save your own from a crop you know to be clean. The Mexican bean beetle does not bother us here but dustings of pyrethrum (a vegetable insecticide) will help. Dust should be applied when plants are damp with dew in early morning.

No account of the legumes would be complete without a mention of their value as stock food. Sweet clover was described earlier, a noble soil builder and fine fodder plant. Even better as fodder is alfalfa which, here in California, can be sown in spring and, as soon as it gets its roots down, needs no further watering. It is necessary to keep it free of weeds at the start, so best sow it in rows 2 feet apart and hoe until it is established. An occasional irrigation helps a lot if the stock graze it heavily. There are, in addition, various clovers (white, red, alsike, ladino) which, seeded in combination with a suitable grass, make ideal permanent pastures or temporary pastures (leas).

Grasses. The grasses are one of the most successful of nature's experiments. It was the development of grasses that ensured the dominance of the mammals over the reptiles during the cooling phase of earth's climate in the Pliocene period. Vast seas of grass on which grazed vast herds of mammals on which preyed smaller groups of mammalian carnivores have been characteristic of earth's biosphere since the Cenozoic era began (about 70 million years ago). This scene was still visible on the Great Plains of this continent as recently as 1850; vast seas of grass, vast herds of buffalo, and tribes of Indians preying on the buffalo. The white man (that arch-destroyer) changed all that. Gone are the buffalo, gone are the Indians. The seas of grass are still there, but it is grass of another species, wheat and corn, the grasses which gave man his civilization.

So man owes a lot to the grass family (Gramineae) as do many other mammals. The only reason I list grasses after legumes is that they never have discovered the trick of teaming up with bacteria to fix nitrogen. But given enough nitrogen they will outgrow legumes and, outside of the forests, are the dominant form of vegetation on the planet earth.

Wheat. Every commune of Whole-earthers should grow its own wheat crop. This plant is so ancient, so sacred, so intimately bound up with man's civilization that no one who values rich impressions should miss the opportunity of watching it grow, holding the ripe ears, rubbing them out in his hands, chewing the fresh grain. He might reflect as he does so that this marvelous grass has sustained men for perhaps 9,000 years and that, nowadays, such is its value, man grows it on 400 million acres of soil, from which he harvests about 250 million tons of grain.

So even though this country grows plenty of wheat (about 34 million tons a year, of which it exports nearly 50%, thereby permanently reducing the fertility of its own soil), it is still worth the Whole-earther's while to grow a plot of his own. We grow a 100 $\times$ 100-foot plot every year and get more than enough wheat to satisfy our needs. The wheat is a part of the rotation, by which I mean we divide our garden into plots 50 $\times$ 100 feet each and put the crops on different plots each year. Wheat needs a clean piece of ground, one relatively free from weeds. So it should follow a row crop like peas or potatoes that has been kept free from weeds. A dirty patch of wheat (one full of weeds) will not yield a good harvest.

We have been in the habit, in October or November of each year, of sowing the wheat broadcast, which requires about 50 pounds of seed for a 100 $\times$ 100-foot plot. The practice of sowing seed broadcast is a very ancient one and involves one of the most archetypal of rhythms, a smooth swing of the arm, a steady rotation of the hand coordinated with the pace of the sower, producing a smooth application of seed. A very biblical exercise

FOOD GROWING 31

—"a sower went forth, and, as he sowed, some seed fell by the wayside." Which is just the trouble with the broadcast method. Too random.

> "One for the pigeon and one for the crow,
> One to wither and one to grow."

So goes the old rhyme. The Whole-earther, therefore, may prefer to confine his biblical technique to broadcasting oats as a cover crop and sow his precious wheat with a small seed drill. This puts the seed into a trench at the right depth (1½ inches), the right distance apart (8 inches) with a foot between rows. This is about the right spacing for a maximum crop which, if the soil is well prepared and fertile, may yield as much as 2 tons per acre. (An acre is 43,560 square feet, a plot about 209 $\times$ 209 feet.)

There are spring wheats and winter wheats, hard wheats and soft wheats. Winter wheat, here in California, can be sown in November, grows through the winter, ripens by June, requires no irrigation. In harsher climates (Montana, for instance) spring wheat is sown early in spring as the soil is fit. A hard red wheat is best for bread making. It contains a lot of gluten, the elastic substance that holds the bubbles of carbon dioxide produced by the yeast. Soft white wheat contains less gluten and more starch, is suitable for crackers, cakes, cookies and pastries but makes heavy bread. Most of the starchy white wheat is grown in Washington, Oregon and California. We are experimenting with Gaines wheat, a dwarf wheat that because of its short stem does not fall over (lodge) in a high wind.

The Whole-earther will harvest his wheat by hand in June or July, when the moisture content of the grain is around 25%. He will, if he uses a sickle, develop an aching back and generate plenty of sweat but have the satisfaction of knowing that he is engaging in a rite that goes clear back to Neolithic times, when crops were cut with sickles made of stone or hard clay. He should use one hand to hold the bunch of stems and, with the sickle, sever the bunch near the ground. The stems are laid neatly together until the right amount has accumulated. This will be bound by his assistant into a sheaf, using either binder twine or the straw itself. Traditionally men wield the sickles, women bind the sheaves. The sheaves are then stacked together like tents (stooks) and left to dry.

When the heads are thoroughly dry, threshing can begin. A Whole-earther interested in doing things the hard way can use a flail to beat the grain out of the heads or engage in a grain dance whereby the heads are stomped under foot until the grain rolls out, splendid exercise and very archetypal but slow. We found an old chaff cutter a satisfactory threshing machine. It broke up the heads, liberated chaff from wheat, was faster than the grain stomp. As for winnowing, a good brisk wind was all that we needed. The chaff blew away, the grain remained. For final separation and cleaning the grain was thrown into a tank of slowly running water. Chaff rose to the surface, floated off. The good grain sank, was spread out to dry, then was stored in sacks in rat-proof bins.

We must admit, when it comes to converting grain into flour, we are not above availing ourselves of electric power. A really honest Whole-earther would probably use wind power or water power or use his arm muscles. There are good hand mills available (try Quaker City Hand Grain Grinder, from Nelson and Sons, P.O. Box 1296, Salt Lake City, Utah 84110). Whole wheat flour which has all the bran left in makes a good rugged loaf, no food for weaklings, nothing like the tasteless white pap sold in supermarkets. For a somewhat less rugged product the bread maker should make one coarse grind, sieve out some of the bran, grind again.

Bread making, that ancient and sacred art, uses the process of fermentation as does wine making and brewing. The yeast mixed with flour and water ferments some of the starch in the dough. Bubbles of carbon dioxide are generated, trapped by the elastic gluten. The bread rises. In the oven the bubbles expand, the gluten is fixed, the bread is baked. There are lots of tricks to making bread (try the *Tassajara Bread Book*, Shambala Publications Inc., 1409 Fifth Street, Berkeley, Calif. 94710).

Corn. Corn, or maize, is so old a friend of man and has become so thoroughly domesticated in the process that it is now absolutely dependent on man for its propagation. The hundreds of seeds on the massive ears have to be separated if they are to grow. We have experimented with three kinds of corn: sweet corn, field corn, Indian corn. The first sowing here in California can be made April 15. For a continuous supply of fresh sweet corn, sowings should be made on May 1 and 15, June 1 and 15, July 1. The largest sowings to give extra corn for freezing should be made on May 1 and 15. These will escape the corn ear worm. Later sowings will be chewed by this creature.

Corn should be sown in drills at a depth of 2 inches, 6 inches between seed, thinned later to 18 inches, 3 feet between rows. Seed corn is generally a hybrid of two different strains because corn, more than most plants, exhibits the phenomenon of hybrid vigor. As it is rather difficult to raise one's own hybrids, corn seed is one of the few seeds we buy rather than save (our preference, Golden Cross Bantam).

Corn is a hungry crop. A plant which grows from a seed to an 11-foot giant in five months is bound to take a lot out of the soil. No other crop will tell the Whole-earther more about the state of his soil than corn. It is often said that "corn speaks" and so it does to one who can understand its language. Small, weak plants with yellowish leaves tell of nitrogen shortage, purple discoloration of leaves and stems of phosphate lack, yellowing of the edge of the leaf of potash lack. The Whole-earther who makes a religion out of "organic" gardening may be able to grow good corn if he applies plenty of manure. One less fanatical will use two applications of 10-10-10, one at sowing, the second when the tassel forms.

Corn is wind-pollinated and should therefore be sown in blocks of not less than 4 rows at a time. We sow rows 50 feet long and we sow 10 rows for the big plantings (using nearly all of it for freezing) and 4 rows for the small. Using immature corn as sweet corn is, to some extent, self-indulgence. A more conscientious Whole-earther would disdain the practice because it uses a valuable crop before it is ripe, at a time when it has developed only a fraction of its nutrients. Fully ripened corn is more nutritious than sweet corn and needs no freezer (drawing its power from environment-polluting utility companies). Even ripened corn is not a good food, however. The protein of corn is not only low in quantity (about 10% as against 15% in wheat), it is also low in quality. It contains almost no tryptophan and little lysine. However, a high-lysine corn has now been developed and peoples who rely largely on this crop as a food source can, by the use of the new strains, at least avoid malnutrition. The high-lysine strain can be obtained from Pioneer Hybrid Seed Corn, Johnston, Iowa.

Rye, oats, barley. These grasses are mainly of interest to those Whole-earthers whose soil is too poor for wheat or too cold for corn. They are the poor man's cereals. Rye will grow where wheat will not. It is the hardiest of all the cereals, grows tall and rank, so much so that it tends to lodge, so should be grown only on poor soil. The grain makes a heavy black bread, food of the peasants throughout Poland and Russia. Rye or mixtures of rye and vetch can be used as a cover crop.

The oat differs from wheat and rye in having a husk that cannot be readily separated from the grain. It needs a special roller mill for its preparation unless one is ready to digest husk and all. Oats have for generations provided the food of Scots Highlanders. They eat it in the form of brose, which is raw oatmeal taken with milk. A rugged diet. One who can survive on brose can survive on anything. Which does not mean one should despise the oat. With its loose floating pannicles it is a beautiful grass, makes a fine cover crop, can be cut for hay, puts fire into the blood of horses and grows fairly well on poor soil, magnificently on rich.

Barley, like the oat, has a grain which is firmly enclosed in its husk. It will ripen early (two months between sowing and harvesting) and requires little moisture, so is cultivated extensively here in California on dry rolling hills east of Napa that will grow nothing else. Barley is very low in protein but, when sprouted, generates a lot of sugar called maltose. This is a major component of malt. The Whole-earther can make his own malt if he grows a crop of barley. Harvest your barley early to get a high concentration of carbohydrate and a low level of protein, steep the kernels in water and spread them out on a concrete floor keeping them moist, raking them over at intervals. As the barley starts to sprout the grains convert their starch into sugar. Dry the malt in an oven, grind it, crush it, extract with hot water. This liquid is *wort*. All one need do to convert the wort into beer is boil it, cool it, add yeast, and flavor with hops. If you boil the fermented wort and condense the vapors you end up with a product called *usquebor* or whisky. You can do similar things with corn, in which case it is called moonshine or mountain dew. Watch out for Revenooers. They take an interest in home distilleries.

Rice. If the Whole-earther lives in the Orient or in certain parts of the United States or Australia he may want to grow this precious grass. He will have plenty of company. The noble community of rice growers is probably the biggest, certainly the wettest and hardest working of all the world's cultivators of cereal crops. Rice is the staple food for more than half of the human biomass. It is by far the most demanding of the cereals. Passionately addicted to water, it has to be grown in paddies, which are shallow ponds surrounded by banks of mud (bunds). The soil in the bottom of the paddy is stirred up by a suitable plow under 2 inches of water. Rice plants raised in a nursery plot are set out in the mud 4 to 16 inches apart. The plants grow with their topmost leaves just above the water so the water level in the paddy is raised until the rice has flowered. Then the level of the water is lowered and, as the seeds form, all the water is drained off. The foliage withers, the crop is cut, threshed and winnowed.

All in all it is a fine experience, especially for one who enjoys sloshing in mud. Here in California one can grow rice easily, given a year-round creek and a reasonably level piece of soil. It is possible to get quite good crops from improved strains of rice such as IR-8. The one thing not to do with your rice after you've grown it is to polish it. This is a crime against a fine food. It removes almost all the proteins, fats and vitamins, including the precious

B_1. So millions of Asiatics who live almost solely on rice suffer from *beriberi* in addition to malnutrition. You cannot be properly nourished on nothing but rice. It is too low in protein. All Whole-earthers living in Asia should therefore practice and demonstrate to others the following procedures: (1) Never polish the rice. (2) Use high-yield varieties and harvest two crops a year. In some areas of Australia, Italy and Spain yields average 2–3 tons per acre. In Asia yields of half a ton or less are common. (3) Use the flooded paddies to raise fish (tilapia) for protein. (4) Introduce a legume into the crop plan (soybeans, for instance). It fertilizes the soil and balances the diet.

Pasture grasses. If you have more land than you need for a garden, use it for pasture. Goats, sheep, cows, chickens and pigs can convert grass into high-protein food. Pasture is rarely good as is. It can be improved by introduction of more nourishing grasses combined with clovers. So great is the variety of climates and soil that the Whole-earther must find out for himself which grasses are best suited to his land. In dry areas he may find it worthwhile to plant millet or sorghum. Rye, grazed when green, makes good forage and you can get a grain crop from it later. The same is true of oats.

Sugar cane. It would be remiss not to mention this big grass from which more than half the world's sugar supply is obtained yearly. But a mention is all it gets. To grow it you must live in a place that stews and soaks (average temperature 70° F, rainfall 60 inches). Most Whole-earthers probably would prefer to rely on sugar beet for sweetening, or, better still, honey.

It is possible in temperate climates to grow sorgo for syrup. Sorgo syrup is also called sorghum molasses. Minnesota and Indian Amber varieties grow in the northern states, Honey, Hodo, Iceberg, White African and Sugar Drip in the south. Four seeds per foot of row, in rows about an inch deep, is about right. Spacing between rows should be 36 to 40 inches. This grass is enormous when full grown, reaching a height of 12 feet, and it requires manure, preferably applied the year before, or 300–500 pounds of 5-10-5 per acre.

To make syrup, remove the tops along with the upper joint or two and strip off the leaves. The stems must then be run through a crusher, which can easily be made from the parts of an old chaff cutter. From a ton of sorgo cane 10 to 15 gallons of syrup may be obtained and an acre of sorgo will give 8 to 10 tons of stripped topped stems. Quite a small area of sorgo will, under favorable conditions, produce all the syrup needed by the family.

Solanums. I mention the solanums (members of the Solanaceae) next because this family contains two of man's favorite foods, the potato and tomato, as well as one of his favorite poisons, tobacco. Like maize all three came from the New World, were unknown in either Europe or Asia until the sixteenth century.

Potatoes. This tuber-producing solanum is surely one of man's best friends. He plants the crop on large areas of the biosphere, harvests from it about 280 million tons of tubers a year (about 80% of this total is water). It is not a particularly good food for, apart from the water, it contains too much starch, not enough protein. It is also plagued by a host of diseases, one of which (late blight) caused the Irish potato famine of 1846–47, in which a million people died and 1.5 million had to emigrate. From this the Whole-earther can draw the obvious conclusion: never rely for your food on a single crop.

Despite its weaknesses the potato, grown on suitable soil, yields more calories per acre than rice, corn or wheat (potatoes yield 4.84 million cal-acre, rice 2.77, corn 2.55, wheat 1.61). The plant will grow on almost any fertile soil except very heavy clay. Best is a sandy loam friable enough to allow the tubers to expand. In California one can plant potatoes as early as February. One can even put them in in November in frost-free, well-drained corners. It is best to buy seed potatoes rather than save one's own. Potatoes are cursed with a host of virus diseases which become worse year by year and are carried over from crop to crop. Seed potatoes produced in Maine or Idaho (or Scotland for the British Isles) are relatively free from these viruses. Potatoes are actually swollen stems and the eyes are buds. Your seed potatoes will go further if you cut them in pieces, at least two eyes to a piece, and allow them to sprout before planting. They are planted 12–15 inches apart with 2 feet between rows after danger of frost is over. Earth is gathered up around the plants to give the small potatoes protection from light and a friable bed in which to expand. If exposed to light the tubers turn green and become mildly poisonous.

Potatoes should be harvested when the skins are firm, dried in a shady place, weighed and stored where the mice won't get at them. A clamp or bank is useful. This involves putting the potatoes in a mound on a well-drained bed of straw mouse-proofed with wire netting, covering with more straw and a layer of earth on the outside (fig. 4). The thickness of the walls of the

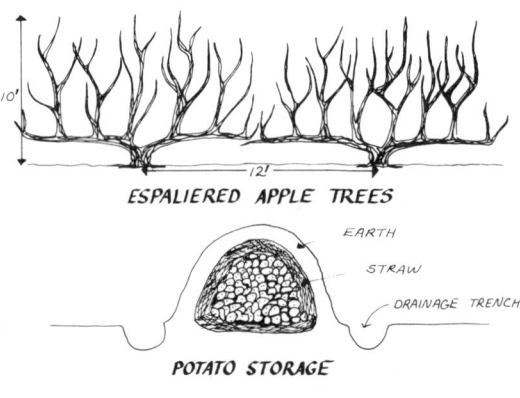

FIGURE 4

clamp depends on the coldness of the winter. The tubers must not be allowed to freeze. Here in California the problem with potato storage is not freezing but early sprouting and sprouts have to be removed at least twice during the winter.

Worst pest of the potato is the late blight caused by the fungus *Phytophthora infestans*. Commercial potato growers spray the plants with various fungicides, of which old-fashioned Bordeaux mixture (a blend of copper sulfate and lime) is as good as any.

The Whole-earther who distrusts sprays in general may prefer to put his faith in resistant varieties. The late blight fungus comes from the central highlands of Mexico in 16 varieties and potatoes resistant to one may succumb to another. Work is in progress to develop a potato resistant to them all. Much depends on the weather. Damp, cool days in June–July are dangerous. Once the disease has appeared it is generally too late to control it.

Tomatoes. These solanums can hardly be classed as a major food crop. They are even more watery than the potato. They are valuable as a base for various sauces, salads and cool drinks high in vitamin C. The seed can be sown in a greenhouse in March, plants set out in May. If you support them on strings or bamboo poles, trim them to one stem, put them 18 inches apart, 3 feet between rows. Or let them sprawl, in which case plant 2 feet apart in the rows, 3 feet between rows. They like sun and plenty of water during the early stages of growth but are very tricky about fertilizer. Potash and phosphate agree with them but an overdose of nitrogen gives huge plants with no fruit. A fleshy variety like San Marzano is good for sauce (Italian style). Our favorite here in California is Ace, large, tasty and long-bearing. It produces right into November.

Cucurbits. This noble family, Cucurbitaceae, includes pumpkins, squashes, cucumbers, gourds and melons. They are scramblers and spreaders. They need room. They tend to megalomania, producing those gigantic fruits that figure in county fairs, giant pumpkins, giant squash. The Whole-earther with a small garden may hesitate to give space to these sprawlers which must be planted about 6 feet apart and do not yield very heavily. The best winter squash are butternut and acorn. The giants, banana squash and the Hubbards, have little to recommend them apart from impressive size. They tend to be as flavorless as blotting paper.

All squash and pumpkins are frost-sensitive and must be sown only when the soil is warm (around May). All have male organs in one kind of flower, female organs in another. They are cross-pollinated by insects and they hybridize freely so, if you wish to save your own seed, grow the different varieties far apart or you may end up with something that is half butternut—half pumpkin and not much use for anything. Pumpkins and winter squash can be stored in a dry cool place (frost-free) on a bed of straw. They last until March out here in California, after which they grow colonies of varicolored molds and rapidly rot.

Brassicas. These plants are among man's oldest friends. They include cabbage, broccoli, cauliflower, kale, Brussels sprouts and kohlrabi. They have been in cultivation since 2500 B.C., were worshipped by the Egyptians, highly esteemed by the Greeks and the Romans. Cato the Elder regarded consumption of large amounts of cabbage as the secret of good health and long life. He lived to the age of ninety so he may have been right.

All the brassicas like a rich soil with plenty of nitrogen and potash, so a heavy dressing of manure is in order or plenty of compost. It is always best to grow the plants from seed in flats (seed boxes) and to transfer the plants when they have three well-developed leaves. Here in California we grow two crops, one to mature in summer, another for the winter. Cabbage (Danish Ballhead or Large Red) can be set out in March, or earlier if the ground is fit. Plants should be set 12 to 18 inches apart, depending on whether you want large cabbages or small, with 2 to 3 feet between rows. Cauliflowers are more frost-sensitive than cabbage so should not be set out until danger of hard frost is over, 18 to 30 inches apart in rows separated 3 feet. Broccoli resembles cauliflower but is easier to grow (these are among the few vegetables in which the part eaten is the flower). Spacing is the same as for cauliflower. Kohlrabi resembles a cross between a cabbage and a radish and plants are set out 6 to 8 inches apart in rows separated 18 inches.

Brussels sprouts are not suitable for spring sowing and should be planted from May to July, 2 to 3 feet apart, 3 feet between rows. This late planting date is suitable for the other brassicas if a winter crop is desired. It is in the cool weather of fall that cauliflowers and broccoli produce the best heads. As for Brussels sprouts they are no good until touched by frost. The summer planting, however, necessitates watering the plants and protective measures against the voracious flea beetle (a small beetle which jumps like a flea and eats holes in the leaves). A dusting with pyrethrum (a natural vegetable insecticide) applied in the early morning when the plants are damp will protect them from this pest.

Cauliflower and broccoli have to be eaten as soon as ready. Overgrown broccoli tastes like a mixture of bird seed and string. But cabbage, in a mild climate, can stay out right through the winter and so can Brussels sprouts. In harsh climates cabbages can be stored (hung up in a cool place) or converted into sauerkraut. Sauerkraut is made by slicing the cabbage, packing it tightly in a barrel and allowing it to ferment. If air is excluded the fermentation stops as soon as certain acids have developed (hence the sour taste). But if air is not excluded the cabbage will decay and the resulting mess will be fit for neither man nor beast.

Sweet potatoes. These plants are fine sources of food for Whole-earthers living in warm climates. They should not be confused with yams. The sweet potato belongs to the morning-glory family, the yam to the genus *Dioscorea*. They have in common that they are climbing plants and produce tubers (the greater yam, *D. alata,* can produce tubers that weigh 60 pounds). Both tubers are rich in carbohydrates but low in protein. It is an overreliance on yams that is mainly responsible for the prevalence of kwashiorkor among children in tropical Africa.

Sweet potatoes do well only in climates where the frost-free season lasts at least 4 months, when warm weather and sunshine extend through the growing season and the nights are warm as well as the days. They also like a lot of water (an average rainfall of 40 inches). Soil should be a light sandy loam and well drained. The roots are set in hot beds of sand warmed to 80° F until they sprout. Sprouts with attached roots are set out in ridged soil (ridges 10–12 inches high) after danger of frost is over. Plants can be set 12–18 inches apart with 3–3½ feet between rows. Cultivation of the rows can continue until the vines meet in the middle.

There are two kinds of sweet potatoes, the soft and the firm.

The soft are more popular in the United States, the most important variety being the Puerto Rico. The firm-fleshed varieties include Big Stem Jersey and Yellow Jersey. New varieties are being introduced by the U.S. Department of Agriculture. Roots should be harvested before frost, cleaned and stored in crates. Sweet potatoes keep best if they are cured after harvesting by keeping them at a temperature of 70–85° F for a period of a few days to 4 weeks (depending on temperature). Best storage temperature is 55–60° F. They can also be stored in banks or clamps like Irish potatoes. It is better to make several small banks than one large one.

4. Minor food crops

Bulbs. The Whole-earther's garden will not be complete without several plants that can hardly be classed as major food sources. Among these are bulbs and roots. Bulbs are really large buds with swollen leaves in which the plant stores food which it uses to produce a seed head in the following year. Many bulbs of the amaryllis or lily families delight the gardener with blossoms in early spring—"daffodils that come before the swallow dare to take the winds of March with beauty." The edible bulbs, however, belong to a genus that does not produce showy flowers. They are all members of the genus *Allium,* of which the onion is the best known.

Onions, leeks, garlic are all old friends of man, familiar for their powerful odor and taste since prehistoric times. There are two ways of growing onions, from seed or from sets. We grow ours from seed, sowing seed in boxes in January or February, setting the plants out in March, 6 inches between plants, 1–1½ feet between rows. Or the seed can be sown directly in the soil as soon as the soil can be worked. The seed is fine and the seed bed must be clean of weeds, firm and well raked. If the soil is wet the good Whole-earther will lay down boards before he walks on it and will firm the soil over the onion seed with the same boards, leaving it level and smooth. Planting onion sets is no problem. Just stick them in the soil 6 inches apart with the tops protruding. Sets have a bad habit of going to seed instead of producing bulbs. They are, however, easier to grow than plants.

Leeks make a winter or spring vegetable here in California. Sow seed in flats in May, plant out in holes so that the base is blanched, and dig them up in March. In colder climates they must be planted in spring, harvested before the soil freezes.

Garlic can be planted in California from October to January and harvested in June and July. It needs cool weather to develop good bulbs.

Onions can best be stored by braiding them by their tops into strings and hanging them up in a cool place. They will keep without sprouting till April. Onion seed can be obtained by planting out good, firm onions in spring and drying the seed heads after the seeds have developed.

Roots. Several plants store their food in swollen roots and use the store to produce a seed head in the following year. The carrot, beet, radish, turnip, parsnip and celeriac all belong to this group. All these plants are grown from seeds. Like the onion they must have a clean well-raked seed bed and must be carefully weeded during the early stages of growth. Two crops can be grown, one for use in June sown as early as the ground can be worked, a second sown in June–July for use in winter. The root crops can be sown in rows 1½–2 feet apart and the plants thinned as they develop. The carrot likes a deep, loose soil and will not do well in heavy clay. Both beets and carrots can be stored during the winter in banks or boxes of sand in a cool place. Radishes and turnips must be used as soon as ready but will give several crops in a growing season. The Whole-earther who really wants to be independent could try growing sugar beet for sugar, but the process of sugar making is rather time- and fuel-consuming. Better to keep bees.

Leaves. Though the cabbage family may provide the major supply of leafy vegetables in the Whole-earther's garden, other edible leaves should also be included. Chard is one of our standbys. It is drought-resistant, does well on heavy clay, can be cut again and again and when too coarse for man can be fed to the goats. The seed can be sown in early spring for a summer crop or in summer for a winter crop. It grows steadily throughout the mild California winter, can be eaten raw or cooked, tastes much like spinach but contains less oxalic acid. The rhubarb chard with its deep red leaves is decorative as well as useful. Sow in rows 18 inches apart, thin plants to about 6 inches. Spinach can be grown in much the same way but is no use in hot weather. It has to be sown as early in spring as possible. Kale is winter-hardy and a fine source of vitamin A, more nourishing than chard. Mustard greens, collards and dandelion can all be used to fill that gap during spring when greens are hard to get.

Good salads, however, are hard to prepare without lettuce, an old friend of man which has been in cultivation for at least 2,500 years. Finest of all for flavor and consistency is the butterhead lettuce (bibb or Boston). It is infinitely preferable to those tasteless bags of water called iceberg lettuce with which California inundates the nation. Bibb lettuces are small and sturdy, can be grown from seed in a greenhouse even during the winter. They can be set out in spring 8 inches apart, 1 foot between rows. The wise Whole-earther, who wishes to pass the lettuce leaf test, will avoid setting out too many at a time and will make fresh sowings every 2 weeks. This applies also to radishes, which go well with the lettuce. A shady, cool spot is desirable if lettuces are to be grown during the hot weather. The large Cos lettuce supplies an abundance of greens but must be sown early (around March). Chinese cabbage, in mild climates, can be set out in October.

No salad is really worth eating without herbs and a herb plot should have place in any Whole-earther's garden. Herbs are aromatic. Their essences in the form of volatile oils add that important ingredient that makes the difference between a dull dish and an interesting one. The leafy herbs sage, rosemary, thyme, savory, basil and marjoram belong to the labiate family and have flowers like those of the dead nettle. A few plants are enough. Thyme, rosemary and sage are perennials so do not need resowing. The umbellates have flower heads like umbrellas and include dill, parsley, anise and fennel, all annuals. Finally, chives, a member of the onion family and a good perennial, can be easily grown as an edging for the garden and will impart a mild onion flavor to salads if chopped fine. Borage, a powerful plant, can be used in salads or as a tea but tends, left to itself, to take over the whole garden.

Celery is worth growing only if a good well-drained, muck soil is available. The plant has a long growing season, needs much water and cool weather. Here in California we sow seed in flats in June, plant out in August. The plants can be spaced 4–10 inches apart with 2 feet between rows and the stems can be blanched by drawing up soil around them after wrapping them in paper. They are resistant to light frost, will stand right through the winter and form a good salad vegetable in February and March.

Mustard and cress is a real standby for any Whole-earther who needs greens during the winter and has no other way of getting them. The seed can be sown on damp cheesecloth in trays and allowed to sprout. It will furnish a continuous supply of greens right through the winter. To replenish seed supply, a crop can be sown in the spring and the seed gathered when ripe.

"Luxury" vegetables. Asparagus, rhubarb and globe artichoke are perennial plants that can serve two functions. They will help hold the soil and will furnish in season some rather tasty tidbits. The true Whole-earther would probably despise such luxuries and not bother with them, especially as asparagus needs constant care and weeding, and artichokes and rhubarb are big plants and tend to take over the whole garden. We grew our asparagus from seed, planted it in rows 2 feet apart, began harvesting after 3 years. The crop is hardly worth the effort. Our globe artichokes grew enormous and were so decorative that they warranted a place on a slope where the soil tended to wash out. Likewise the rhubarb.

Green peppers and eggplants are worth a place in the garden, can be sown in the greenhouse in flats and planted out in May, 2 feet between plants, 2 feet between rows. Out here they go on bearing right into November but the eggplants are the favorite food of flea beetles and, unless liberally dusted with pyrethrum, will be eaten to shreds.

Mushrooms are another luxury crop which the true Whole-earther will probably despise. However, if one happens to have a cellar and access to plenty of horseshit, growing this fungus (*Psaliota campestris*) is worth a try. The horseshit must be carefully composted (heat to at least 160° F), turned and composted again, then loaded into boxes of a convenient size. The careful mushroom grower must be sure that his compost has stopped heating before he places the spawn in it or his spawn will be cooked. The perfect temperature for a mushroom house is around 60° F. After placing his mushroom spawn in the compost the grower covers the mushroom bed with a thin layer of sterilized casing soil. All in all a lot of work for a fungus that is nutritionally almost worthless but the compost is useful and can be spread on the garden after the mushrooms have finished with it.

5. Fruits and nuts

The words fruit and vegetable are used very loosely. Botanically speaking, a fruit is the swollen ovary of a plant and may be either fleshy like a tomato or hard and dry like a hazel nut. The seed is contained in the fruit. For some reason we call tomatoes, squashes, cucumbers and peppers vegetables although they are actually fruits. The difference, I suppose, depends on whether they are eaten with the meat course or the dessert. Most fruits also grow on trees but a melon is called a fruit though it does not grow on a tree, and a strawberry is called a fruit though it does not grow on a tree and is not (botanically speaking) a fruit.

However, we need not get technical. We will simply follow the vernacular. The fruits of the earth, as we know them in temperate climates, belong mainly to the noble family of the rose. These fruits fall into three big groups. To the first group belong the pomes, including the apple, pear and quince; to the second belong the drupes, including plums, apricots, nectarines and peaches; to the third belong a number of soft fruits loosely called berries, including the strawberry, raspberry, blackberry, loganberry and boysenberry. We will now consider these jewels of the rose family.

Apples are intertwined with man's oldest myths, and when painters wish to portray that celebrated exchange which resulted in Adam and Eve's expulsion from the Garden of Eden it is in the form of an apple that they show the fateful fruit. But the apple has so many virtues that it should be a symbol of salvation rather than of downfall. It has remarkable keeping qualities, is fairly nutritious and can be preserved in the form of applesauce or apple butter.

A good way of growing apples, and other fruit, if space is limited is the modified espalier (fig. 4). The trees are planted 18 feet apart at an angle facing south and are trained to give two main stems, one facing south, one north, running horizontally. The upright branches from these horizontal main stems can be pruned to be less than 10 feet high. The horizontals can finally be grafted to those of the next tree. Thus a continuous hedge is formed running north and south, easy to spray and pick, taking a minimum of space.

In pruning his apple hedge the good Whole-earther must realize that most apples bear their fruit on short spurs. These are easily recognized and must be preserved. The leafy shoots that take the form of long whips must be removed or the hedge will rapidly become a jungle. Careful pruning and cultivation around the base of the tree, a light application of manure in the fall and a spraying during winter with lime sulfur are all that are needed to keep the hedge in good health. Here in California the Red Delicious and Golden Delicious do well and Gravensteins are grown commercially to produce thousands of gallons of applesauce. We also have remnants of an old orchard with a variety of apple so obscure that even the experts cannot name it. It produces better applesauce than the Gravensteins.

Pears, like apples, can be grown as a hedge. They need the same sort of care and pruning (preservation of spurs, removal of vegetative shoots). They are somewhat more difficult to grow than apples, being prone to a disease called fireblight which makes the shoots turn black as if burned. Such shoots should be cut out well below the blackened area and burned and the pruning shears should be disinfected.

Peaches, plums, apricots and nectarines bear fruits called drupes, which have a fleshy outer layer and a hard inner stone within which is the seed. The trees can also be hedged and grow luxuriously, in fact will turn into a complete jungle unless pruned vigorously. We grow the Santa Rosa plum and a Japanese variety called Satsuma, both very healthy. Peach leaf curl is a plague of peach trees here but can be partly controlled by Bordeaux mixture, a relatively harmless spray. Those who grow apricots and can the fruit should crack open the pits and add the kernels. They taste like almonds and are highly nutritious.

Blackberries, raspberries, loganberries and boysenberries all have the same characteristic. They bear fruit on long shoots that were produced the year before. After bearing fruit these shoots die and should be removed or the berry patch will become impenetrable. These plants need support and protection from birds.

Strawberries are perennials that spread by sending out long runners which produce new plants at their ends. It is best to plant them about 18 inches apart and let the runners fill in the gaps. What with birds attacking from above and gophers attacking from below, we find nothing short of a fortress of wire netting will protect the berries. They need much watering also and a light, friable soil. They are subject to root rot.

The almond is another member of the rose family that yields edible material, in this case the seed being eaten and the rest of the fruit rejected. The tree is such an early bloomer (February) that only in very mild climates will it produce nuts. We have an almond tree and it sets fruit but the squirrels always get it before we do.

Fruits not belonging to the rose family include grapes, figs, citrus, avocadoes, such tropical forms as mangoes and papayas, and such fruit-bearing bushes as currants and gooseberries. The Whole-earther living in temperate climates will, if he has any devotion to Dionysus and warm enough summers, almost certainly want to grow grapes. Not to do so is to miss one of life's richer experiences. We here in northern California are in the very middle of the grape growing region. All around us new vineyards are being planted at a feverish rate and we ourselves have 3 acres in luscious Golden Chasselas that bear huge clusters of greenish-golden fruit.

The grape belt is a fairly broad one. It stretches from Spain to the shores of the Caspian and extends into northern Africa and to Israel. In North America it includes parts of New York State, much of California, and can be extended into Oregon, Washington and British Columbia. Australia, South Africa and South America also have their grape belts.

Vines need a freely draining soil. They grow well on steep hillsides, as the vineyards of the Rhineland attest. They like a climate with an average annual temperature of 60° F. They are easily damaged by frosts in the spring, are susceptible to mildew and phylloxera and need a lot of care. Also, out here in California they are utterly ravaged by the voracious deer unless the grower puts up a fence 7 feet high. Even so, a small vineyard of some choice wine variety (Pinot Noir, Pinot Chardonnay, Cabernet Sauvignon) is well worth having simply for the experience it offers in the ancient art of wine making. Or, if table grapes are required, a tasty variety like Thompsons Seedless or Muscat can be grown.

Vines are best planted in deep holes December to March, distance apart being regulated by pruning plans. Vines can be trained on a trellis and cane-pruned, or tied to a stake and spur-pruned. Which method is used depends on the variety. American varieties such as Concord are usually cane-pruned. European varieties are spur-pruned. The American varieties are characterized by their slip skin and foxy flavor. The best wine grapes are European varieties.

To avoid damage by phylloxera, grapes are grafted to special resistant stocks. To control mildew, a plague of European grapes, the vines must be dusted with sulfur when the shoots are 6 inches long and again at intervals depending on the weather. Grapes should be gathered for wine making when the sugar content is over 18%, preferably about 20%. For details of wine making, see **Food Storing and Processing**, section 7.

Citrus fruits, the lemon, orange, grapefruit and citron, all belong to the family Rutaceae. Common to all of them is the presence of strongly scented aromatic oils. They are highly frost-sensitive but here in the California hills do quite well. We harvest a huge crop of lemons from two lemon trees and plenty of oranges from the thick-skinned navel orange. They require very little care and, unlike our other fruit trees, are not attacked by deer.

We have dreamed for some time of planting an avocado tree, the fruit of which offers a substitute for butter. Perched as we are in the thermal zone where frosts are rare we might just be able to persuade one of the hardier varieties to bear fruit. Such varieties are Mexicola, Jalna, Duke. They need well-drained soil, frost protection, a sheltered corner and water during the dry weather.

Of nut trees the English walnut is an obvious choice for anyone having available a deep, rich soil. The tree needs deep soil moisture and protection from water at the base of the trunk with a layer of stone. Pecans have similar soil requirements but need even deeper soils (6–10 feet deep, and well drained).

For those living in the tropics the most helpful of all nut trees is the coconut. This extraordinary tree has the power to migrate over thousands of miles of ocean, the nut buoyed up by the dense mass of fiber in which it is embedded. The coconut grows on the most lonely of Pacific atolls, is almost the sole source of plant food for those who live on the low islands, furnishing also great quantities of oil, plus sap which can be fermented to form *toddy* or *arrack*.

Another generous tropical palm is the date palm, capable, if properly treated, of producing as much as 100 pounds of dates per tree per year. Date palms are carefully rooted from offshoots that arise at the foot of the trunk. They have separate male and female flowers and have to be hand-pollinated for full production. The date is so rich in sugar that, if properly dried, it will keep indefinitely. It is thus the staple food of the wandering

Bedouin, who manages to survive on dates and camel milk and little else.

6. Animal foods

Whether or not he should keep animals is a question every Whole-earther must decide for himself. Helen and Scott Nearing, those pillars of the Alternate Society, decided right from the start when they fled from New York City to Vermont that keeping livestock was out. They gave their reasons in *Living the Good Life* (Schocken, 1970). Keeping animals is a form of slavery which enchains both the animal and its keeper. The Nearings refused to eat animal bodies and refused to enslave either themselves or the beasts. It must be added, however, that they had no children and therefore felt no impulse to offer the little ones a glass of milk or a boiled egg, fine sources of protein for growing bodies.

It makes sense to keep animals only if one wants animal protein and wants it badly enough to be willing to pay the price which keeping livestock imposes—namely, that one must first buy the creatures (often a considerable cash outlay), then house them, feed them when natural food is short (another cash outlay), and always run the risk of their contracting some disease and dying, or, in the case of sheep, being killed by dogs, which are the bane of the sheepman's existence, at least around here. The conversion of plant protein into animal protein is wasteful. It is commonly stated that about 10 pounds of corn are needed to produce 1 pound of beef. This is an exaggeration, but there is always loss on conversion. Probably the most efficient converter is the trout and some other fish, hence the importance of aquaculture (see section 8). The pig will generate 1 pound of pig per 3 pounds of feed during its most rapid stage of growth and 1 pound per 5 pounds of feed later. It does especially well on high-lysine corn. A young steer (600 pounds) needs 17.5 pounds of food to make 2.6 pounds of beast. Beef cattle, however, can use food materials that are useless for man. They have, in a special compartment of the stomach (rumen), vast swarms of microorganisms capable of digesting cellulose and generating amino acids even from such unpromising materials as urea and ammonium salts. The goat, another ruminant, is also an efficient converter and can live, but not thrive, where other animals would starve.

However, if one wants milk, neither goat nor cow will provide it in large amounts unless fed supplements. Milk is a complex mixture. Cows' milk contains about 12.7% solids, of which 3.3% is protein, 4% is fat and nearly 5% is lactose (milk sugar). Ewes' milk is much more concentrated (6.5% fat, 6.3% protein). Milk from the Jersey cow has more fat than that of the Holstein (5.58% versus 3.59%). A mature cow giving 50 pounds of milk a day and weighing 1,000 pounds will need about 3 pounds of protein and 24 pounds of digestible nutrients a day. A good pasture containing 50% legumes and 50% grass will provide much of the needed raw material for milk production. But low-protein roughage such as oat hay needs to be supplemented with a concentrate containing 17–18% digestible protein. The concentrate, however, has to be purchased and it makes little sense to spend more on the concentrate than the milk is worth.

The true Whole-earther, whether he raises poultry, cows, sheep, goats or even rabbits, will be wise if he considers the advantages of "folding" his flock. Sheep folding has been practiced in England for centuries and the method can be extended to other livestock. It consists in improving pastures by sowing mixtures of clover and grass or growing special crops (turnips or kale, for example). Then the area is divided into enclosures with temporary fencing. The animals are released into one of these enclosures. When they have eaten the feed they are moved to the second, then to the third, then the fourth. In this way the animals manure the ground and gather their own food. I have often thought of using the system with rabbits. The rabbit is a good source of protein but if you keep it in a cage and feed it pellets the rabbit meat hardly pays for the food. Better to fold your rabbit flock, using movable pens, movable shelters for the night and improved pasture. You get the rabbit shit straight on the ground and raise your herd with a minimum of bought food. The same can be done with chickens if you clip their wings. The stock is kept healthier this way and the garden benefits.

Sheep are by far the least trouble of all livestock. Out here in California they graze on the dry hills and seem to survive with little supplemental feeding. Though the wool is hardly worth the shearing if one sells it, it does offer amateur spinners and weavers and rug makers a chance to try their hand at an ancient craft and meat eaters can stock the freezer with the corpses of lambs. Also, sheep do not wreck orchards and gardens as do goats and can be kept in place with a minimum of fencing.

The Whole-earther who prefers to rely on eggs for animal protein should remember that chickens, turkeys, ducks and geese lay eggs that contain 70–74% water, 12–14% protein and 12–14% fat. The eggs are a more concentrated food than milk. To produce an egg a chicken must have nearly double the amount of food that it needs simply to stay alive. The use of concentrated bought foods fed to small flocks of hens in smelly little enclosures is a procedure of questionable economic value. If costs are taken into account the eggs so produced are often absurdly expensive. Better let your chickens scratch for their own food, feed them cheap grain and accept fewer eggs per bird.

7. Crop layout

After he has familiarized himself with the requirements of various crops the Whole-earther must work out a program. Correct land use demands crop rotation. It is not a good idea to grow the same crop year after year on the same piece of ground. The layout shown in fig. 5 may seem excessively elaborate and may involve more fencing than the average Whole-earther is willing to install. It is, however, based on sound principles. The idea behind it is to alternate temporary pastures with vegetable crops and to obtain the maximum amount of both animal and vegetable food from the smallest possible area of land. It is true, as was mentioned earlier, that any procedure involving the conversion of vegetable to animal material is wasteful. But if you are going to raise animals for food, whether they be cows, chickens, goats or rabbits, you will come closer to the goal of being self-supporting if you feed the beasts on food you have raised on your own land than if you buy animal food at today's exorbitant prices. It is good agricultural practice to grow special crops for the animals and let them eat the crop where it stands. The fertility of the soil is improved and the farmer is saved the work of harvesting the crop. In addition to grass-clover mixtures and

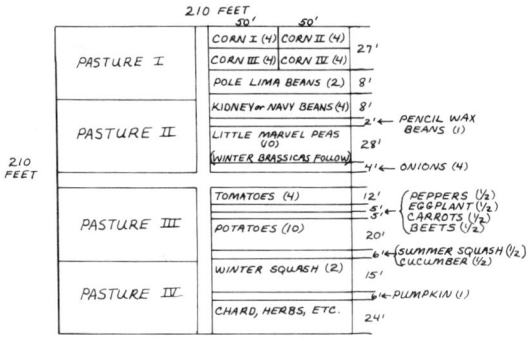

CROP LAYOUT

FIGURE 5

alfalfa, such crops as kale, mangels and turnips can be grown for the stock. If the Whole-earther does not want the trouble of putting up extra fencing these crops can be brought in and fed to the animals in the barn.

The organization of the vegetable garden shown in fig. 5 is, of course, only one of many possibilities. The local climate, the layout of the land, distribution of sun and shade all affect the crop plan. The following points should be remembered:

(1) Avoid putting tall plants such as corn on the south side of short plants unless you wish to shade those plants.

(2) Keep plants that will remain in the soil through the winter in a separate area. A winter garden of such hardy vegetables as winter peas, fava beans, brassicas, chard and Chinese cabbage requires well-drained soil, and a southern exposure if possible.

(3) Use plants such as rhubarb and asparagus which remain in the soil for several years to hold the soil in areas where erosion is a problem.

FOOD GROWING

(4) A more detailed example of crop layout than that shown in fig. 5 is given in table I, for a garden 100 × 150 feet. Spacing of rows can be facilitated by use of the adjustable spacer shown in fig. 6. A garden this size will provide food for one biological

TABLE I

Sample vegetable garden, 100 × 150 feet (⅓ acre)
Vegetables, with length of rows in feet given in parentheses

Total width of rows, feet	
	North end
5	Rhubarb (30) Asparagus (120)
6	Double row lima beans (150)
3	Early sweet corn (40) Midseason sweet corn (40) Late sweet corn (70)
3	" " "
3	" " "
3	" " "
3	Pole beans (75) Early tomatoes (75)
4	Tomatoes (150)
2	Lettuce (20) Spinach (100) Radishes followed by Lettuce (30)
2	Beets followed by Turnips (75) Carrots (75)
2	Carrots (150)
2	Onion plants (75) Sets (50)
3	Snap beans followed by Spinach (150)
3	Chard (30) Parsley (10) Parsnips (50) Chinese cabbage (60)
3	Cauliflower (40) Brussels sprouts (30) Broccoli (80)
3	Eggplant (50) Peppers (50) Celery (50)
3	Peas (150) followed by Cabbage
3	" "
2	Potatoes (150)
2	"
2	"
2	"
6	Summer squash (40) Cucumbers (110)
7	Winter squash (150)
7	Melons (50) Winter squash (150)
	South end

56 FOOD

unit (male, female and two offspring), with the exception of wheat or rice, essential to complete the diet.

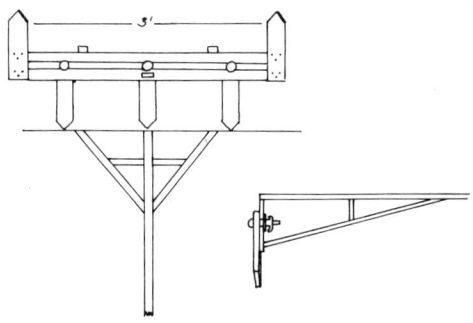

SPACER FOR SOWING

FIGURE 6

8. Aquaculture

There are compelling reasons why the true Whole-earther should turn his attention to aquaculture for part of his food. First, 75% of the planet's surface is covered by water. Second, water, if correctly used, will produce more animal protein than an equal area of dry land.

There are two kinds of aquaculture, salt-water and fresh-water. The salt-water variety involves owning or renting some coastal property. Fresh-water aquaculture involves owning a stream or lake or making a pond.

Fresh-water aquaculture. In his book *Fish Culture* (Faber and Faber, London, 1971) C. F. Hickling described some of the reasons why it is worth engaging in this form of aquaculture: "Fish culture can use land too poor to be useful agriculturally, and water with qualities unsuitable for agriculture or irrigation. Swampy land, undrainable for agriculture, makes very good fish pond sites. Land which may be toxic to plants because of a very acid reaction may be suitable for fish ponds. In Israel brackish water with as much as 8,000 to 10,000 parts per million salt are used for fish culture. Major fish farming industries are based on the changing brackish waters of estuaries."

For the true Whole-earther, who is often forced by lack of means to buy land of poor quality, this statement may be of special interest.

Fish culture in fresh water is governed mainly by temperature. Such fish as catfish and tilapia require water between 70 and 80° F for maximum growth. Trout, however, will hardly survive in water above 70° F and do best in water around 56–58° F.

Fish ponds. A number of points should be borne in mind by any Whole-earther who contemplates making a fish pond:

(1) Try to excavate your pond so that it has a flat bottom surrounded by banks. The best average depth of water for fish is 3 feet and the only reason for making the pond deeper is to prevent it from drying out in the dry season.

(2) Always try to locate your pond on a site high enough so that it can be drained by gravity. Draining is desirable not only because it facilitates harvesting the fish but also because it increases the fertility of the pond and improves oxygenation of the water. Ideally, the fish grower should have two or three separate ponds, which should be dried out in rotation every 2 years. A good crop can be grown on the bottom mud.

(3) If possible, ensure a water flow through your pond. This can be done by diverting part of the flow from a small creek to pass through the pond. Catfish, even with intensive feeding, will generate only 1 ton of fish per acre per year in water that is static. If 200 gallons of water per hour can be directed through the pond, production can be raised to 3,500 pounds per year. A still larger flow will enable one to harvest 5,000 pounds per acre.

(4) Quality of the water will determine the success of the fish growing venture just as quality of soil determines the outcome of an agricultural enterprise. The pH (degree of acidity or alkalinity) of the water is important. It can safely range from 6 to 9. At an

58   FOOD

acid pH of about 4 most fish die and water more alkaline than 9.5 inhibits growth. Calcareous water with high levels of dissolved lime are more productive than non-calcareous. To determine alkaline reserve, take 100 milliliters of water, add three drops of methyl orange indicator. If it turns pink the water is too acid. If it turns pale yellow, add 0.1 normal hydrochloric acid drop by drop until it turns pink. Water of medium productivity requires 8–30 drops. The best alkaline reserve is indicated by 30–75 drops. To correct acidity, scatter powdered limestone on the water.

(5) A spillway must be constructed to enable one to empty the pond and to take off excess water during heavy rain. The most convenient form which will serve both purposes is shown in fig. 7. This spillway must be constructed before the walls of the

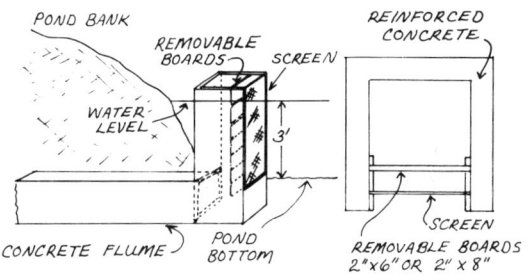

POND CONSTRUCTION

FIGURE 7

dam are raised and must have sufficient diameter to take the runoff from a heavy rain storm. The screen prevents loss of fish when the dam is drained. The boards enable you to regulate the depth of water. Before the season of heavy rain some boards can be removed to lower the water level and accommodate the new inflow.

(6) It is distressing to discover, after much time and effort has been expended on pond construction, that the water leaks out

faster than it enters. This frequently happens in newly constructed ponds. It generally indicates that the banks were not packed down firmly enough while they were being built. The leaks often seal themselves after the first heavy rains. It is possible in small ponds to seal them with a sheet of plastic (black polyethylene is best but must be leached for several days before fish are introduced). The plastic can be covered with a layer of mud or sand. Dressings of bentonite (a fine clay) will also help seal the leaks in ponds.

(7) Oxygen level in pond water is very important. Fish depend on the dissolved oxygen (D.O.) in water and its level decides which fish you can grow. The D.O. level in freshly filled ponds may be very low so no pond should be stocked with fish immediately. Allow about 40 days for oxygen levels to build up.

(8) To produce the maximum crop of fish a fish pond must be fertilized. Many fish farms in Java are run in conjunction with pig sties, the manure being transferred directly to the ponds. Poultry manure and human manure are also used but the employment of untreated human sewage is objectionable on account of the danger of spreading disease. In Munich, Germany, sewage effluent is used to support a crop of carp. Organic matter, such as manure, must be added cautiously to the pond, preferably when the temperature of the water is low. The decay of the organic material uses oxygen and may therefore asphyxiate the fish. The warmer the water, the less oxygen it can hold in solution. Phosphate is the most important fish pond fertilizer and can be applied at the rate of 25–30 kilograms of phosphorus pentoxide (P_2O_5) per acre. Potash is rarely deficient in fish ponds.

(9) Draining the pond is the only efficient way of harvesting fish. Seining is much less effective. Using a large seine and dragging it through the pond 10 times, Hickling found that the haul on the tenth pass was not much lower than the haul on the first. Ponds can be completely cleared of fish by the use of derris root, the active principle of which is rotenone. This substance is both a fish poison and an insecticide but it is harmless to humans. A dose calculated to give 6 parts derris per million of water will be sufficient and will not make the water toxic for more than 5 days.

The derris should be spread evenly over the surface of the pond. The fish will be stupefied and float to the surface. They can be collected with dip nets. Care must be taken to ensure that the pond is not overflowing when the derris is applied as the fish poison may contaminate other water. Fish stunned with derris can be quickly revived by putting them in fresh water.

A freshly made slurry of quick lime is also a fish poison, and can be used to disinfect a pond after it has been drained. The quick lime (calcium oxide) becomes converted into calcium carbonate, which is beneficial and aids the growth of the fish.

(10) Food for fish in ponds should be placed in definite stations near the bank and marked off with floats. Regular inspection of these feeding stations will show whether fish are eating all the food. Household scraps, spoiled grain, even fresh grass can be fed to fish if the right combination of species is used (for example, grass carp that will eat fresh grass, leaves and vegetable refuse). Maize, wheat and lupin seed should be soaked before feeding to fish.

Fish grow better if the feed is supplied frequently, so mechanical food dispensers are coming into use in fish farming. There are even demand dispensers on which the fish learns to push a button, thereby releasing food whenever the fish needs it. Fish must be watched to see if they are using all the food. In very hot weather care must be taken not to overfeed the fish. The excess food will rot and lower the oxygen content of the water. Hot and windless days will put fish off their feed.

The fish farmer must know approximately how fast his fish are growing as food supply must be proportional to their weight. A sample of fish should be removed from time to time. This can best be done with a small fish trap or cast net. Fish should be disturbed as little as possible as they lose weight when frightened. It is usual to feed 5% of the weight of the fish per day but this applies only to foods of high food value. Less nutritious foods must be supplied in larger amounts.

Raceways. A raceway is a channel through which water flows at a rapid rate. It is particularly suitable for raising fish having a high oxygen need such as trout. The dimensions of the raceway will depend on the amount of water available. Raceways used for high-density catfish culture at the Skidaway Institute of Oceanography in Georgia consist of a series of segments 10 feet wide and 100 feet long containing water 3–4 feet deep. There are 25 of these segments. Water from a 3-acre pond flows by gravity through the first segment and over a 1-foot-drop aeration weir into segment two, and so on. At the end water is pumped back to the pond. A flow rate of 0.05 gallon per minute per cubic foot of raceway makes it possible to raise fish at a density of six fish per cubic foot. The method is thus well suited to high-density fish culture but the layout is expensive compared with the fish pond. It may be necessary to line raceways with plastic or rocks to prevent erosion of banks by the rapidly flowing water.

Trout. These are cold-water fish of the salmon family. Ideal water temperature is around 58° F. Brown trout can stand water up to 70° F and rainbow trout up to 75° F but these high temperatures are bad for the fish. Although raceways with rapid water flow are best for trout the fish can be raised in ponds. A very successful trout growing operation has been started in Denmark using earthen ponds. Rainbow trout have been raised and show very fast growth in ponds fed by sewage effluent at Munich. Rainbow trout do not need as much oxygen as brown trout or brook trout. Trout are carnivorous fish and voracious feeders. In the United States, where a considerable trout-growing operation is located on the Snake River near Buhl, Idaho, the trout are fed specially prepared pellets. In Denmark they are fed on trash fish. The true Whole-earther would probably prefer to set up a more self-contained biosystem by using leaves and manure to grow worms, feeding the worms to the trout, putting the spent compost on the garden. If he runs out of worms he can buy trout chow. The fish are amazingly efficient food converters and, according to J. E. Bardach (*The Status and Potential of Aquaculture,* vol. II, P.B. 177768, from Clearinghouse, Springfield, Va. 22151), can convert 1½ pounds of food into a pound of fish. Special strains of trout with rapid growth rates have been developed by Lauren P. Donaldson of the University of Washington.

Although trout can be raised from eggs the fish farmer would do best to buy fingerlings from a hatchery. They cost here in California about $30 per 100 (rainbows). They can be raised to a satisfactory size in 2 years or less, depending on the feeding schedule. Details of feeding rate can be obtained from manufacturers of trout chow.

Catfish. These are warm-water fish and their maximum growth occurs in water at 80° F. It is quite possible, if one is willing to take the trouble, to raise one's own fish from eggs. The fish spawn when the water is 80° F. Pairs of selected brood fish are placed in 5 × 10-foot pens of hardware cloth or chicken wire located around the edges of the pond. The fish need sheltered places in which to build their nests. These can be wooden boxes partly filled with gravel, nail kegs or crockery jars. Several inches of water must cover the tops of the nests. Fish weighing 1–4 pounds produce 4,000 eggs per pound body weight. Eggs hatch in 5–10 days at 70–80° F.

It takes a year to bring catfish from egg to fingerling size. A second year is required to bring the fish up to ¾ to 1 pound. Those unwilling to try to raise fish from eggs can buy them as fingerlings of various sizes. The cost here in California (1973) is about $15 for 50 half-inch fingerlings. Catfish can be raised in raceways at a density of two to six fish per cubic foot of water (see J. W. Andrews, ed., *High Density Fish Culture,* Skidaway Institute of Oceanography, Skidaway Island, Savannah, Ga., 1970). The equipment needed for such intensive rearing is expensive and probably beyond the means of the average Whole-earther. In the rich bottom lands of the Mississippi, which are the center of catfish culture in the United States, ponds are used which are commonly no bigger than 1 acre. Channel, blue and white catfish are raised in these ponds and the yields, if the pond is unfertilized, will be around 200 pounds per acre. If the pond is fertilized or the fish are fed, the yield can be raised to 1,000 pounds per acre. A flow-through of water can increase yields to as much as 5,000 pounds per acre.

Quite good yields can be obtained from unfertilized ponds by using a mixture of fish species. Without much labor one can harvest from an acre of water 200 pounds of catfish, 100–300 pounds of buffalo fish and 50–60 pounds of bass per year.

Carp. This noble fish, famed for its wisdom and longevity, has been cultured in carp ponds for centuries. In old England every monastery had its carp ponds and the monks fed the carp and the carp fed the monks and both grew fat in the process. In favor of these fish are the following qualities: They breed readily, stand low temperatures, do well in stagnant water. Near Munich they are raised in ponds into which the partially treated sewage of the city is led. The sewage promotes growth of algae, the algae are eaten by various worms, the worms are eaten by carp, the carp are eaten by the inhabitants of Munich. A perfect example of recycling. Carp can also be trained to feed themselves on cheap foodstuffs compounded into pellets like those used in trout culture. A carp presses a target plate in the water and a pellet is released. The carp fancier can take his pick of a variety of species, from the highly decorative oriental carp to the silver carp or grass carp. The silver carp from the Amur River is particularly popular in the Soviet Union. It is herbivorous, has delicious tender flesh and few bones. In Israel the carp is grown along with mullet and tilapia and the yield of fish in these mixed cultures increases because the entire volume of the pond is used as a three-dimensional growing space.

Most Whole-earthers will probably prefer the simplest form of aquaculture involving the use of some animal wastes to fertilize the water and raise a mixed culture of bass, bluegills, carp and catfish with a few bullfrogs thrown in to make music at night. In this way the Whole-earther can go fishing whenever he feels like it, catch his supper with rod and reel or use a seine net if his skill as an angler is insufficient. The trouble with most farm ponds is that they are not fished enough and the resultant overcrowding produces masses of fish too small to be worth eating.

Ocean aquaculture. This type of aquaculture can be divided into three parts: (1) culture of sea plants, (2) of invertebrates, and (3) of fish or marine animals (ocean fish culture).

Sea plants. The plants of the sea take two forms: single-celled algae which make up the *phytoplankton* and which float freely in the upper layer of the ocean; and the larger plants (seaweeds) that are usually anchored to rocks and grow in shallow water relatively close to the shore.

The culture of one-celled algae whether in sea water or in fresh is a large subject (for more information, see the letter by Pat Patterson, *Last Whole Earth Catalog*, p. 59). There are those who have suggested that such algae cultures will be humanity's last resort to save it from mass starvation. Probably, when things get that bad, this ill-starred hominid will prefer to become extinct. There is, however, a very practical reason for growing one-celled marine algae. Such algae are the natural food of oysters and abalone and their culture will be described in connection with the growing of these invertebrates. Here we are mainly concerned with the culture of seaweed, particularly the kind called *nori* in Japanese and laver in English. Laver culture in Japan is quite technical, for the Japanese are surely the world's leading aquaculturalists. The secret involves gathering special spores (carpospores) and letting them grow on oyster shells. This is done at the marine laboratory in tanks $6 \times 8 \times 3$ feet holding 250 strings of 10 oyster shell collectors. The carpospores grow on the shells as a dark red incrustation and release monospores in the autumn. At this time the fishermen who grow laver come to the laboratory with special nets which they immerse in the tank, paying the equivalent of $1 per net. The monospores settle on the net, which is then set out in an estuary. The seaweed grows during the winter and is harvested from January until April.

The seaweed is highly nutritious and digestible, the algae protein constituting 30 to 50% of the dry weight. It is said to have a nutritional value comparable to beef. One acre of intensively cultivated seaweed will bring in $3,000. Unfortunately, the Japanese, besides being the world's best aquaculturalists, are also the world's worst polluters and are progressively rendering their estuarine waters incapable of growing anything. But a true Whole-earther with a relatively unpolluted estuary within reach might try laver culture (see M. Kuogi, "Recent laver culture in Japan," *Fishing News International,* July–September 1963; or particularly for invertebrate and algae culture, see John H. Ryther et al., *The Status and Potential of Aquaculture,* vol. I, P.B. 177767, Clearinghouse, Springfield, Va. 22151).

Invertebrates. Oyster culture and abalone culture represent two methods of growing first-class animal protein that might recommend themselves to any Whole-earther with access to open water. It is not very difficult to raise oysters and the project can be combined with sewage purification. Such a scheme has been worked out by John H. Ryther, who runs two "farms" at Woods Hole Oceanographic Institution. In the first farm, treated sewage is used to culture phytoplankton in carboys exposed to light. This operation can be done in a greenhouse or with artificial light. The algae cleanse the sewage by removing much ammonium, nitrate and phosphate, though they will not grow in sewage polluted by toxic industrial wastes. The sewage is mixed with aerated sea water pumped in from the ocean. Water carrying the algae is then pumped into the second farm, consisting of large tanks in which are suspended strings of scallop shells covered in oyster spat. The oysters grow on these shells using the algae as food. Such oysters can be harvested and fed to people. Another example of recycling.

It is also possible to raise oysters in ponds of sea water as is being done on Long Island. The baby oysters can be suspended on strings of scallop shells or on plastic netting called Netron (made by Du Pont).

By suspending his oysters in this way from rafts or buoys the grower saves them from various hazards that threaten oysters grown on the bottom (fig. 8). This method of culture by suspension is used very successfully in Hiroshima Bay by the Japanese, who harvest the world's largest oyster crops: 58,000 kilograms per hectare as opposed to 10 kilograms per hectare from the public oyster grounds in the United States (see J. E. Bardach, "Aquaculture," *Science,* vol. 161, pp. 1098–1106, 1968).

The Whole-earther can have fun and learn a bit about marine

biology if he raises his own seed oysters. Oysters can be forced to spawn by introducing them into warm sea water (about 23° C) and the oyster spat can be collected on strings of scallop shells on wires suspended in the tank from cross pieces or on strips of Netron. Less enterprising Whole-earthers may prefer to buy their baby oysters. They can get them from Japan or from the Lummi Indians of Lummi Island, Wash., who have gone into aquaculture in a big way and are specialists in oyster growing.

The whole process of growing marine invertebrates, including shrimps, lobsters, crabs, clams, mussels and abalone, is in an

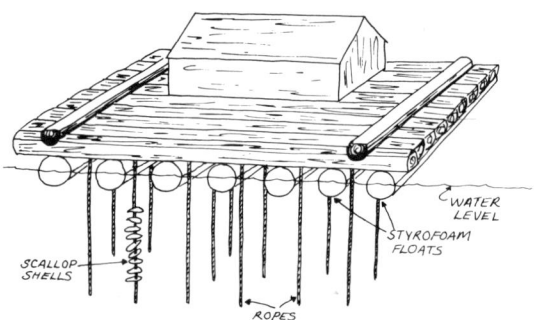

RAFT FOR OYSTER OR MUSSEL CULTURE

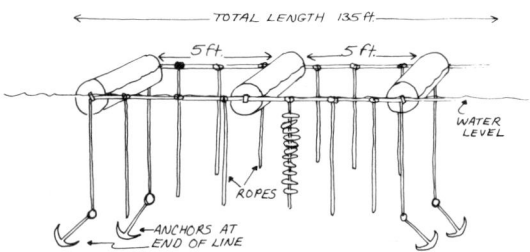

LONG LINE METHOD OF OYSTER CULTURE

METHODS OF OYSTER OR MUSSEL CULTURE

FIGURE 8

embryonic stage in the United States. It is of special interest to Whole-earthers who live in such states as Maine that have ample shore lines. Writing in *The National Fisherman* (February and March 1971) Dr. Robert L. Dow, Marine Research Director, Maine Department of Sea and Shore Fisheries, describes the present state of the art. He sees Maine aquaculture as being worth far more than the state's existing industries. The great threat to this wholesome activity is, as usual, industrial pollution of coastal and estuarine waters.

Ocean fish culture. The chief apostle of ocean fish culture is Dr. Lauren P. Donaldson of the University of Washington. His specialty is the rainbow trout, of which a sea-going variety is known as the steelhead. A trout and salmon breeding program which has been in progress for more than 30 years has resulted in the development of some truly remarkable fish. Whereas a wild rainbow trout in a lake will reach a weight of only 200 grams in 18 months one of the new hybrids may grow to an astonishing 3,000 grams in the same period. Selective breeding can be used to produce similar improvement in salmon (salmon and trout belong to the same family of fish, the Salmonidae).

The idea underlying fish farming in the ocean is to release specially bred fish to forage in the sea ("the big pasture") but to keep them under control, just as a rancher keeps his cattle under control. This has been done in places as far apart as Norway and Tasmania. Concerning this method of farming the sea Dr. Donaldson has this to say: "In Tasmania, for instance, the culturists are using estuarial water and farming rainbow trout and brown trout in the sea in the type of floating enclosures that are so popular off the coast and along the fjords of Norway. The Tasmanians do not have to go to a great expense, but they do have their animals completely under control at all times. Their food supply is natural food plus other food. The fish are readily available for marketing any day you want them. They simply say that it is much better to keep the fish under control than to hunt all over to find them as you do in the regular commercial fisheries."

FOOD GATHERING

Our contemporary prophets of doom vary as to the vision they see of life after the Ultimate Power Failure. Some envisage a return to about the year 1850, some see a return to the late Neolithic, some to the Paleolithic. No doubt, if our mad militarists are ever permitted to use their enormous stockpile of instruments of destruction, a return to the Stone Age will indeed occur. It is, in any case, a good idea to be prepared for such an eventuality.

The Stone Age is not very far away. In a corner of my California garden, close to a creek, in the shade of great pepperwoods and redwoods, I come upon arrowheads and chips of obsidian that speak of the ghosts of departed indians who camped on that spot before the white man robbed them of their land. And to the west, in the Great Basin around Pyramid Lake, the Indians lived in the Stone Age so recently that Margaret M. Wheat could portray their way of life with photographs and give accurate descriptions of their methods of survival in a very hostile environment (*Survival Arts of the Primitive Paiutes,* University of Nevada Press, Reno, 1967).

Anyone who wishes to sample the way of life of our Stone Age ancestors can do so any time he wishes by simply going into a wilderness area with nothing but his bare hands and seeing what happens. In this way he can treat himself to a foretaste of what will occur when the fuel runs out, the walls fall down, the freeways are empty and one turns on the switch and nothing happens. The bold spirit who decides to try this experiment would be well advised first to study a book by Larry Dean Olsen (*Outdoor Survival Skills,* Brigham Young University Press, Provo, Utah, 1967).

Larry Olsen certainly does things the hard way. He takes into the wilderness only the clothes he stands up in. He very rightly remarks, "The challenge of vaulting from the Space Age into the Stone Age may prove too great a shock for some people, and life is often lost because of inability to adjust." The truth of this statement was proved again and again during World War II, when castaways in small boats often died even when they had adequate provisions. They died, it appears, of shock, or simply lost the will to live. That condition of spiritual disorganization loosely known as panic kills more surely than hunger. The first law for one entering the wilderness (or stranded there by accident) is "Keep your cool."

It was to train young people to keep cool and stay alive that the Outward Bound School was originally created in Scotland. Outward Bound schools now exist in America (Outward Bound, Inc., Andover, Mass. 01810). There is also the National Outdoor Leadership School (Paul Petzoldt, Director, Lander, Wyo. 82520), which specializes in training in the art of living in the mountain wilderness (specifically the Wind River Range, Wyo.).

Surviving in the wild depends on finding food and shelter. Shelter building will be described in Chapter II, gathering of food will be dealt with here.

1. Hunting and trapping

Hunting with primitive weapons (spear, bow and arrow, sling or bolas) is a far cry from hunting with a high power rifle with telescopic sights. First the hunter must make his weapon. This calls for skill in the Stone Age art of pressure flaking. The natural glasses, agate, obsidian or flint, are best for flaking. Strike at an angle with a hammer stone to remove flakes from the glass, shape the thin blade by pressure, holding it against the heel of the hand and protecting the hand with a leather pad. The flaker, a piece of horn, should be pressed firmly against the stone blade to remove from the undersurface a flake of the required size. It takes practice to produce a good arrowhead but once the technique is mastered some very pretty stone tips can be produced.

Stone axes are produced more easily than arrowheads or spear tips. All that is needed is a hammer stone, which is used to peck away at the stone being shaped to produce an edge, and a groove by means of which the axe head can be attached to the haft. The axe head should be held in the hand while being worked, not placed on a hard surface. Olsen states that a good axe head can be turned out in about an hour. It can be hafted by heating a slender willow, wrapping it twice around the axe head and locking it in place with suitable fibers (such as those of the nettle).

Arrowheads can be attached to a hardwood shaft (chokecherry, rose, currant or willow). The wood must be straightened and carefully dried, scraped down and smoothed with a chip of stone. It is economical to use the hardwood as a foreshaft only; the foreshaft can be 6 to 8 inches long and inserted into a reed to give a total length of 24 inches. Straight reeds are often easier to obtain than straight pieces of hardwood. To attach the stone tip to the foreshaft a notch is made deep enough to receive the point, which is bound in place with sinew and glued with natural resin or pitch. Feathers are split and attached to either side of the shaft, preferably with glue but if necessary with fiber.

To make a bow a long, smooth, straight branch about the size of a man's wrist should be split with a wooden wedge and shaped and scraped on the split side. The back should be left unworked. Ash, juniper, chokecherry or willow can be used. The bow should be about 44 inches long. Its ends can be recurved by greasing the ends with fat, heating them over hot coals, bending and holding in position until cool. Bowstrings are best made from sinews, strands of which can be twisted together to give any desired length of string. The sinew string is finished by stretching it between two points, rubbing with saliva to smooth rough spots and drying in this position.

For one who is less of a purist than Larry Olsen the best of all primitive hunting weapons is the cross bow. This formidable weapon can be made from automobile leaf springs and will pull at several hundred pounds. Those reluctant to make one can buy one ready-made (*Last Whole Earth Catalog,* p. 278). The wise Whole-earther, anticipating the Ultimate Power Failure, might do well to keep one of these handy. They should be used with care, being powerful weapons and not legal in some parts of these United States.

The *atlatl,* or spear thrower, is another primitive weapon that can be of value to one willing to master its use. The *atlatl* is about 2 feet long and 2 inches wide. At one end it has a prong to fit into the hollow end of the spear, at the other it has two loops through which the fingers of the thrower are passed. The spear should be 5 to 6 feet long with a foreshaft of hardwood tipped with stone. Two feathers tied to the butt will help balance the spear in flight. The spear is thrown overhanded with a sweeping motion and the extra length of the *atlatl* gives it a higher velocity.

The boomerang, blow gun, bolas and sling are four more primitive weapons that may help the hunter to survive in the wilds. The weapons can be made or purchased. Purchasing them is again contrary to the strict, "start at the Stone Age" doctrine of Larry Olsen but the fact is that unless one had a lot of practice with these weapons one's chance of killing game with them is almost zero. A hunting boomerang can be obtained from Col. John M. Gerrish, 4409 S.W. Parkview Lane, Portland, Ore. Blow guns are made by Survival Research Co., 71 Ridge Crescent, Manhasset, N.Y. 11030, or Wham-O Manufacturing Co., 835 E. El Monte St., San Gabriel, Calif. 91778. In case you want to use arrow poison the active ingredient is tubocurare. It will paralyze the game without poisoning the person who eats the game, since it acts only through a wound. Keep out of reach of children!

Hunting with primitive weapons demands patience and intelligence. Here in California the deer are so numerous that they form the natural prey of the hunter. To kill with a spear or arrow it is necessary either to strike the heart or one of the large blood vessels. To do this the hunter must place himself in the right position and at the right distance. To move closer to a deer the hunter must step forward only when the animal lowers its head to graze. Deer can be ambushed, especially at water holes in the evening.

Trapping is a more effective way of catching animals than hunting. The deadfall is a trap designed to kill or immobilize the game. It consists of a rock or log supported by three sticks, forming a figure 4, one of which may be baited. The slightest movement of the baited stick causes the whole structure to collapse,

72　FOOD

pinning the game to the ground (fig. 9). A snare is a loop of wire or fiber placed across an animal trail in such a way that it will encircle the animal's neck. It is connected to a spring pole which, released by the prey's struggles, lifts the animal off the ground. A sunken trap is a hole dug in the trail, covered with leaves and deep enough to prevent the prey from escaping. The taking of a game animal with traps is forbidden in the United States except during an extreme emergency when one is lost in the wilderness.

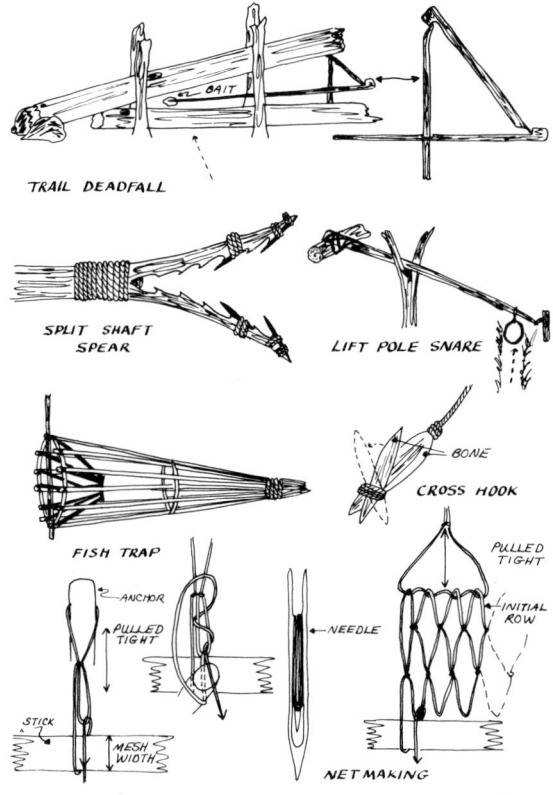

TRAPS AND NETS

FIGURE 9

Mice and other small rodents can be prepared for food by removing skin and viscera, pulverizing bones and adding to a stewpot or roasting. Frogs, lizards and snakes should be skinned and roasted or boiled. Grasshoppers, grubs and ants can be roasted on coals or added to soups and stews. Roast rattlesnake is quite a delicacy. To trap the animal immobilize it with a forked stick behind the head and decapitate it promptly.

Euell Gibbons offers useful tips in *Stalking the Wild Asparagus* (David McKay Co., New York, 1962) in a chapter entitled "How about the Meat Course?" If one is willing to overcome prejudice, says this prince of foragers, one can dine very well off opossum, raccoon, squirrel, muskrat, woodchuck, porcupine, frog, turtle, crayfish and armadillo. Once he even shot and ate a bobcat and found the meat clear, white, tender and delicate. He recommends removing four little pear-shaped glands that are found in the small of the back and under each foreleg on the opossum, raccoon, woodchuck and porcupine. By placing the cleaned animal for 24 to 48 hours in cold water to which half a cup of salt and half a cup of vinegar have been added, the flavor can be much improved.

Crayfish and turtles are both readily available, especially in the east, and the humble bluegill swarms in many ponds. Boneless filets can be made from the bluegills and the skins removed. The filets can be shaken in a paper bag with flour, salt and pepper and fried in hot fat. Even wild carp, a fish apt to taste like a mudpie, can be rendered palatable if skinned; the filets are removed from the backbone with the fingers and dropped into hot deep fat.

2. Fishing and shore foraging

The best place in which to go primitive and stay alive is the coast. Our ancestors knew this and the enormous piles of oyster and other shells that they accumulated give evidence of their fondness for seafood. Euell Gibbons, who forages more for fun than survival, has described how easy it is, without any special equipment, to gather cockles, mussels, clams and edible seaweeds, provided one knows where to look and what to look for (*Stalking the Blue-Eyed Scallop,* David McKay Co., New York, 1964).

The commonest, most easily gathered food organism both on the east coast and the west is the mussel. Few people seem to realize that this mollusk is good eating. It is available in quantity and nothing more is needed to gather it than a low tide, knowledge of where it grows, a tire iron and a sack. It can be steamed open, the meat removed and fried. The acorn barnacles clinging to the shells are also tasty. (Mussels, oysters and clams should not be gathered near populated areas, unless the gatherer knows that the shellfish are safe. Californian mussels are generally quarantined from May to December, during which months they are apt to be poisonous.)

Clams of various sorts and sizes, scallops, oysters, periwinkles, cockles, piddocks, whelks, limpets, crabs, abalone—all these are to be had on the coast and all are edible. For vegetables there are four varieties of seaweed, the laver already described, Irish moss, edible kelp and sea lettuce. The forager can find orach, samphire, sea rocket and sorrel growing on the shore. There is also the beach pea and strand wheat, beach plum and woundwort. No one should starve on the coast.

Nets. The net can be used for trapping game as well as fish. The Paiutes used it to trap rabbits. To make a net one must have cordage. There are three wild plants that yield good fiber: the stinging nettle, the milkweed and the dogbane (also called Indian hemp). To extract the fiber from these plants, soak them in water for a week, dry, and then break, roll or gently pound with a round stone. The fiber can be cleaned by combing through one's fingers. To convert the fibers to cordage, hold two strips of fiber in the left hand between thumb and forefinger, twist strip #1 clockwise with the right hand, lay the twisted strand over strand #2, then repeat the process. Add more fibers to both strands when the ends are reached.

From the cord, nets can be made. First, prepare from a flat piece of wood a netting needle as shown in fig. 9. The needle must be narrower than the mesh of the net to be made. Make a spacer from another piece of flat wood twice the size of the desired mesh. Make a loop in the end of the cord, put it over a firm support, make a second loop about 6 inches from the first. Place the cord over the spacing piece and hold with the thumb (left hand). With the right hand, from below put the needle through the loop and pull it up to the edge of the spacer. Hold with the thumb. Throw the cord over the left hand and from below pass the needle through gap #1 and then gap #2. Pull the cord, not releasing thumb pressure until the knot is tight. This gives mesh #1. Slip out the spacing piece and make mesh #2. Make enough meshes to give a net of the desired length. Gather up the meshes on a loop of cord. Start working back along the row of meshes, adding a second row. The length of the net can be increased or decreased and in this way a net of any shape can be made. Of course, it is a lot easier to buy fish netting (try Netcraft Fishing Tackle, 3101 Sylvania, Toledo, Ohio 43613) but more interesting to make your own.

The net can be used as a seine, held up by floats, the lower edge sunk with weights. Or it can be stretched on an A frame and used as a dip net. Such nets are used in California to catch surf smelt that run in on the breakers at night. Or nets can be used to channel the fish into fish traps.

Fish traps. Use willows if available. Split and shape a willow frame to give an opening 2×2 feet (fig. 9). Space the wands lengthwise to give the body of the trap and bind together at the ends. Spacing of wands depends on the size of fish desired. Use shorter, flexible wands to make the entrance. Place in a stream, facing the direction of migrating fish. A barrier of willow rods or netting will direct the fish into the trap. A good eel trap can be made with netting and two loops of heavy wire.

Small boat fishing. For the ocean a kayak is the best and safest small boat. We use fiber glass and resin to make a 15-foot kayak with 36-inch beam and a large enough cockpit (36 inches) to accommodate two people or give the fisherman room to land his catch. An inflatable kayak-canoe can be bought from Sears

Roebuck for about $90 (1972 price). These boats are safe but hard to handle in high wind and rough water. They can be made safer if the owner adds a spray cover. The back-to-the-Stone-Age purist can hollow a canoe from a log and stabilize it with an outrigger (the Polynesian double canoe was made in this way; it took its builders safely from Tahiti to New Zealand). Primitive boats, somewhat damp and not very safe, can be made from bundles of reeds (used by Paiutes, ancient Egyptians and fishermen of Lake Titicaca) or a raft can be constructed from 50-gallon drums or driftwood. A determined fisherman can always get out on the water (see Chapter V, Watercraft).

Angling. Catching fish with hook and line depends on using the right bait, the right hook, and knowing where to fish. Rock fish off the California coast are a reliable source of food. Anyone with a kayak, a small tent, a good supply of hooks and a frying pan could live on the coast indefinitely on these tasty fish and have lots of time left for contemplation. I speak from experience. I use for fishing an 8–10-ounce weight to which is attached a triple 6/0 hook with a "jig rig" above it of three hand-tied red and yellow lures. The triple hook is baited with fish guts, or octopus which I frequently find in the stomachs of fish I catch. The art, of course, is to know the reefs and kelp beds. "Fish where there are fish."

3. Wild plants

A forager who knows his way around can live off wild food plants, provided he does not make the experiment in dead of winter on terrain covered with several feet of snow. Wild food plants can be divided into three groups: those having edible seeds or fruits, having edible roots or tubers, having edible leaves or stems (fig. 10).

Seeds and fruits. The *white oak, red oak, black oak* and *live oak* produce edible acorns. They were widely used as food by the Indians, who crushed them, steeped them in water for a day to get rid of the tannin, roasted them or ground them into flour from which they made a kind of bread. *Shagbark, mocker-*

nut and *shellbark hickory* all bear edible nuts. *Hazelnuts* and black walnuts need no description. *Pinenuts* were the favorite food of the Paiutes. Their attitude toward this food can be instructive for a Whole-earther: "When we come to a pinenut place we talk to the ground and the mountain and everything. We ask to feel good and strong. We ask for cool breeze to sleep at night. The pinenuts belong to the mountain so we ask the mountain for some of its pinenuts to take home and eat. The water is the mountain's juice. It comes out of the mountain, so we

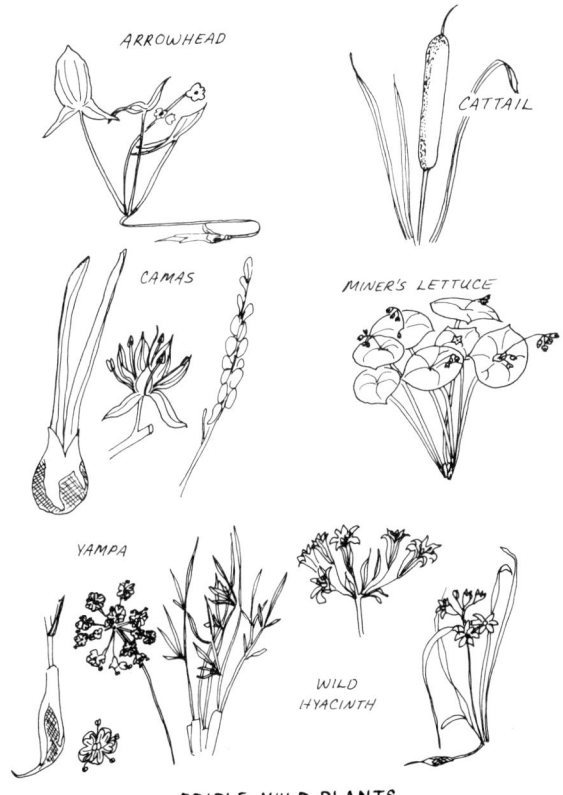

EDIBLE WILD PLANTS

FIGURE 10

ask the mountain for some of its juice to make us feel good and happy." Pinenuts were gathered in the fall, roasted by tossing with hot coals in willow baskets, gently crushed with a stone to break the shells, winnowed to remove the shells, reroasted with coals and ground into flour with a *mano* and *metate*.

Amaranth is an annual herb resembling goosefoot. Seeds can be stripped from the stalk and winnowed for a delicious grain. *Goosefoot* can be treated similarly. The young tops of these plants make good pot herbs. *Mustard* grows wild in many places and the seeds are edible though pungent. *Sunflower*, the small wild variety, has edible seeds. The Indians used the seed as a baby food. *Wild rice,* an important edible seed, is available around ponds and lakes in Minnesota and Michigan. The grain is ¾ inch long and almost black. To gather it the women would paddle out in a canoe into the rice beds, bend the plant heads over the canoe with one hand and strike them with the other. The ripe grain would fall into the canoe. It was threshed by being trampled with the feet and winnowed.

Mesquite, or *honeypod,* will provide food for foragers in the desert regions of New Mexico, Arizona and California. The bean-shaped pods contain sugar and the seeds are also edible. The meal which can be made by drying and grinding the pods and their contents has been called the most nutritious of all bread stuffs. The *honey locust* is another tree that produces sweet, edible pods. It resembles the St. John's Bread, a tree of the dry regions with a range from Spain to Israel. It is thought that the "locusts" eaten with wild honey by John the Baptist were not the insects of this name but the dry, sweet pods of the locust tree.

There are several other wild legumes that have edible seeds. Among them are the *tepary bean,* the *buffalo pea,* the *black medic* and a variety of vetches.

For fruits the forager can try the red berries of the *manzanita,* from which a tasty jelly can be made. The *ground cherry,* with its bladder-like pods, provides a tasty fruit with good keeping qualities. *Red rosehips* are an excellent source of vitamin C and fruits of the related *thornapple* or *hawthorn* (*Crataegus*) can be ground and made into cakes. The Indians mixed them with powdered jerky to make pemmican. Fruits of the *chokecherry* can also be gathered to make pemmican. They contain a cyanogenic poison in the seed and should be crushed and cooked before use to destroy the toxic material.

Roots and tubers. *Arrowhead,* or *wapatoo,* is a water plant easily identified by its leaves, shaped like large arrowheads. The tubers are borne on stalks in the mud. Indian women, using a canoe 14 feet long and 2 feet wide, hung onto the canoe and let themselves into the water breast-deep and fished for wapatoo tubers with their toes. Once freed from the mud the tubers would rise to the surface. "In this manner, these patient females remain in the water for several hours, even in the depth of winter." Any Whole-earther with a farm pond can plant the arrowhead around it and gather the tubers. They can be cooked fresh or dried.

Wild hyacinth in the east and the *camass* in the west provided edible bulbs which were much relished by the Indians. The two plants both belong to the genus *Quamasia* and are related to the onion. The bulbs can be cooked by boiling or roasting or by the Indian method. This involved digging a hole, lining it with stones, building a fire and raking out the embers. The hole was lined with green leaves, filled with camass bulbs, covered with more green leaves and a layer of earth and allowed to steam for a day and a night. It is also recorded that the bulbs, when boiled in water until the water evaporated, formed a very good molasses used by the Indians on festive occasions.

The forager who uses camass must be very careful not to confuse it with the death camass (*Zygadenus*), which has small white flowers, whereas edible camass has large blue ones. The death camass (also called star lily) is very poisonous. The *soap plant* (*chlorogalum* or *amole*) is another plant that has bulbs which were roasted and eaten by the Indians. The uncooked bulbs have lather-producing properties and, crushed, can be used to stupefy fish. The soap plant has white flowers and must not be confused with the death camass.

Bulbs of the *sego,* or *Mariposa lily,* were also steamed in pits and eaten by the Indians of Utah. The Mormons in their first years in Utah consumed these bulbs in great quantities and rewarded the plant by making it the state flower.

Jack-in-the-pulpit, or *Indian turnip,* is an example of a plant whose roots are edible only after treatment. The bulbs of this plant must be boiled and afterward thoroughly dried, then they can be ground into meal and baked into cakes or used for gruel after the Indian fashion.

The *groundnut,* or *bog potato,* is a twining vine with milky juice which grows in thickets and damp places from New Brunswick, Canada, to Florida and west to Kansas. It is considered one of the very best of the wild foods. The sweet tubers may be eaten raw or roasted. Asa Gray, the botanist, gave it as his opinion that, had civilization started in America, this would have been the first edible tuber to be cultivated. Any Whole-earther with a swampy piece of soil could experiment with the domestication of this legume.

Prairie apple, or *Indian breadroot,* is another wild legume which might be worth domesticating. It grows on prairies and high plains from Manitoba and Texas west to the Rocky Mountains. Its root was a favorite item of diet of the Sioux, was peeled and eaten raw or boiled or roasted in the campfire.

From British Columbia to Southern California the forager can find the *squawroot,* or *ipo* or *yampa,* a member of the carrot family related to caraway. It is generally considered the finest food of the northwest Indians. The fleshy roots were dug by the Indians in spring or early summer, washed and trampled under bare feet to remove the brown skin. The roots could be cooked as a vegetable and served with venison or could be dried, ground into flour and baked in cakes.

Salsify, or *goatsbeard,* is a cultivated plant that grows wild in California. It has blue flowers that form conspicuous seed heads with fluffy, umbrella-like parachutes that enable the seeds to float away on the wind. The first-year roots are edible when cooked and taste like steamed oysters. *Indian potato* is a small plant with slender, divided leaflets growing from a small bulb, one of the tastiest roots in the west. It grows in meadows and mountainsides at higher altitudes. *Spring beauty* with its paired opposite leaves and tiny pink flowers also has a small edible tuber and can be an important source of food.

Largest of all the wild edible roots is that of the *wild potato vine,* or *mecha-meck*. This is a member of the morning-glory family and a close relative of the sweet potato. It is found in dry soil from Connecticut south to Florida. The top of the plant is a climbing vine which dies back each autumn but the root is perennial and grows enormous. In *Edible Wild Plants,* Macmillan, 1966, Oliver P. Medsger describes one dug up in New Jersey that weighed more than 30 pounds. The *bush morning-glory,* which grows from South Dakota to New Mexico, has an even larger root than the wild potato. It can be roasted but has a slightly bitter flavor.

Stems and leaves. *Cattail* is a large rush growing by the side of ponds and rivers and the stems and leaves, as well as other parts, are of great value to the forager. The roots, peeled and dried, can be made into flour for bread. The young shoots can be eaten raw or boiled like asparagus, the pollen can be used as flour for bread. The green head before the pollen appears can be boiled and eaten like corn on the cob, and, when dry, the heads can be burned and the tiny roasted seeds eaten. Moreover, the down makes fine insulation and the leaves can be woven into sandals, baskets, blankets and ponchos.

Miner's lettuce, with fleshy disc-shaped leaves that encircle the stem, is one of the best California salad plants. The streams in spring are also loaded with *watercress* and in some ponds the *marsh marigold* grows, the leaves of which can be cooked and eaten like spinach. Leaves of *dandelion, chicory, stinging nettle,* young *milkweed* and *purslane* can all be cooked and eaten if gathered at the right time. Even the young leaves of the malodorous *skunk cabbage* can be used if the water in which the leaves are boiled is changed several times and the leaves are well seasoned with butter, pepper and salt. Young *bracken* fronds (fiddleheads), before they unfold, can be boiled and eaten like asparagus. The creeping root stocks of bracken are also edible.

4. Water

No amount of food will keep a man alive if he does not have water. We are concerned here not with water-getting for domestic purposes, a subject dealt with in Chapter III, but with emergency situations in which water-getting is a life-or-death affair. Such situations arise for the person lost at sea and for the survival-in-the-wilds enthusiast who has chosen a desert to survive in.

Fresh water can be made from sea water by distillation, using solar heat. A still can be made if one has a sheet of plastic, a cup or can and a plastic tube to use as a drinking straw. The same type of still can be made by the desert survivalist. (On the shore, a hole is dug until water fills the bottom.) In the desert, the survivalist must dig a 3-foot hole in soil at the bottom of a gulley or in an old river bed. The can is then placed in the hole. The plastic sheet is stretched over the hole and held in place with sand or dirt. A stone or other weight placed in the middle of the sheet weighs it down until it comes to within 2 inches of the can. Water will condense on the plastic and drip into the can. The drinking tube is placed in the can and extends outside the still. "Two of these stills in operation even in the driest deserts will produce enough water for one person each day" (Larry Olsen). Slices of cactus placed in the pit will increase the amount of water.

Useful amounts of water can be obtained by mopping up the dew in the early morning or spreading cloths to catch dew and wringing them out before the sun dries them. Tapping trees (maples, birch, aspen and white pine) will yield drinkable amounts of water from February through April. Drinking urine or sea water won't help much but a diluted sea water (one part sea to three parts fresh) can be handled by human kidneys.

5. Fire

Modern man generates fire so easily that he can hardly realize how difficult it is to create this aid and comfort when deprived of matches. As long as the sun shines and he has a good lens the problem is not serious. Merely focus the image of the sun on a heap of tinder and soon the welcome plume of smoke arises. The reflector of a flashlight will function as a lens. Put the material to be lit on a small stick passing through the hole where the bulb usually fits until it reaches the focal point in front of the reflector and starts to smoke. But suppose you have no lens or reflector? Then you must either use flint and steel, a bow drill, a hand drill or a rubbing stick.

The flint-and-steel method is not too difficult if the survivalist has thoughtfully provided himself with a semi-functioning cigarette lighter (lacking fuel). Joe Smith of Hood River, Ore. (*Last Whole Earth Catalog*, p. 273), declares that the lighter will supply plenty of sparks and that very fine steel wool is the best tinder ever invented. But cigarette lighters and steel wool are not commonly found in the wilderness. A lump of quartz, agate or flint may be. Strike the stone with the closed end of a pocket knife or any piece of steel. Catch the spark, preferably in a small piece of cloth previously partly burned and bedded in a nest of dry plant fiber. When the spark catches on the cloth, fold the tinder bundle around it and blow on it until it bursts into flame.

A bow drill has four parts: fire board, drill, socket and bow. Hold the fire board in place with the right foot. Put the drill into a depression near the edge of the fire board. Loop the string of the bow once round the drill (the bowstring is best made from a strip of buckskin). Grease the socket and put it on top of the drill. Put the fire board on a piece of bark or wood and rotate the drill in the depression until a pile of dust forms. When there is plenty of smoke and the dust starts to turn black, lift the drill and lightly fan the pile of dust. When the pile begins to glow, add tinder and blow into flame. The hand drill and rubbing stick operate on a similar principle but require much more effort. The bow drill is by far the most practical method of making fire.

To maintain a fire source, use the Indian fire bundle, which is a long bundle of bark like a giant cigar about 2 feet long. Place a hot coal at one end and let the bundle smolder. It can be carried in the hand or in one's belt and will hold a live spark for 6–12 hours.

FOOD STORING AND PROCESSING

1. Salting, smoking, drying

One of mankind's oldest problems is how to store food. Paleolithic man, having with great exertion felled a mammoth deer or any large item of game, stuffed himself with as much meat as he could eat for as long as the meat was eatable and left the rest to rot. However, three methods of storing meat and fish were discovered early in man's history and have been used ever since. These are salting, smoking and drying, often used in combination.

Fish. The fish is gutted, split lengthwise, sprinkled with salt and placed flesh side up on a rack in the sun until dry. Smoking fish (or meat) requires a smoke box or smoke house (fig. 11). The size of the box depends on the amount to be smoked. I use a simple metal box with two trays placed over a rock fireplace. For fuel any hardwood can be used. Rotten wood is preferable because it does not tend to burst into flame. Rotten apple or willow is good. Never use resinous wood like pine or spruce.

The fish should be split and fish over 2 pounds should be fileted (fig. 11). The skin must be left on to hold the flesh together but the scales should be removed, except in species like lingcod and cabezon in which they do not tend to become detached. The time required for smoking depends on the thickness of the filet. A thick filet may require 24 hours or even longer. Mussels and oysters can also be smoked and dried. They make tasty snacks.

Meat. Cut meat into chunks about an inch wide and then convert the chunk into a strip about a half to a quarter of an inch thick. Such strips should be hung in the sun for a couple of days to dry. Salt and spices can be added to the jerky to give it more flavor but they are not essential. Small animals and birds can be dried whole after skinning and eviscerating. The back must be cracked between the legs and a stick inserted to hold the body open. The animals can be laid out on rocks in the sun, protected

FOOD STORING AND PROCESSING   85

by cheesecloth to keep off the flies. When thoroughly dry they can be pounded to crush the bones and left another day in the sun to dry the bone marrow.

Pemmican can be made from jerky by pounding the dried meat with dried berries of chokecherry or hawthorne. Melted suet is then mixed with the berries and the mixture is rolled into small balls and stored in the cleaned intestine of a large animal. If the intestine sack is tied shut and sealed with hot suet, pemmican prepared in this way can be kept for months.

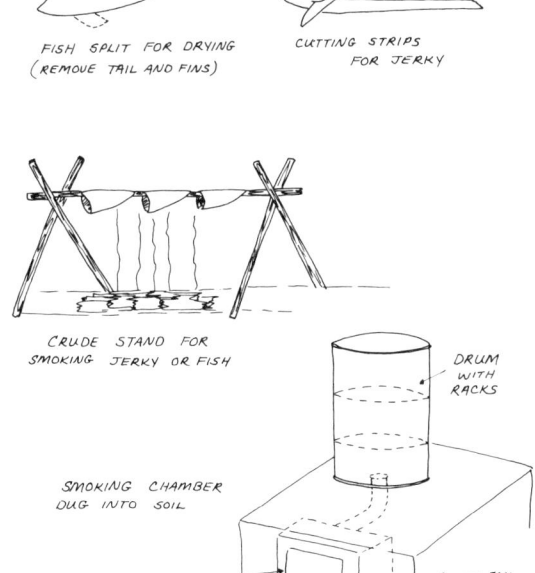

SMOKING AND DRYING MEAT OR FISH

FIGURE 11

Fruits and vegetables. Drying of fruits and vegetables is simple and, if certain precautions are taken, a palatable product can be obtained. These points should be remembered: The more finely vegetables are sliced, the more rapidly they dry. Immersing vegetables in boiling water long enough to inactivate enzymes (about 2 minutes) will improve flavor but may leach out vitamins and sugar. Direct sunlight is destructive and better results may be obtained on trays in shade as long as plenty of air circulates over the tray. Sulfuring the product (exposing to sulfur dioxide from burning sulfur in a suitable box) improves keeping quality, bleaches the product, but is probably somewhat destructive to some of the components. If you store dried fruits and vegetables in sealed containers they should be inspected from time to time. Unless the product is really dry it will tend to go moldy. It may keep better in a cloth sack hung close to the ceiling where the air is warm. Green beans can be threaded on strings and hung up to dry.

Those Whole-earthers unable to rely on the sun as a source of heat can easily build a dryer for fruits and vegetables. Such a dryer is nothing more than a warm-air oven with a sufficient flow of air to carry off the moisture. The amount of fuel involved, however, might be more profitably used to can or bottle the material than to dry it. Of course, the Stone Age purist will want nothing to do with cans or bottles as Stone Age man could not make these containers. But until we are pushed back all the way to the Stone Age we may as well make use of such conveniences. The true Whole-earther, however, will always bear in mind that cans are made from the earth's steadily diminishing supply of metal and are not reusable. Pint or quart jars made from glass are reusable and therefore to be preferred.

2. Canning and bottling

Bottling food will be successful if a few facts are borne in mind. Heat kills the molds and bacteria that cause food spoilage. It also cooks the food, inactivates enzymes, destroys some vitamin C and may impair flavor. But no amount of heat will preserve the food if the product is imperfectly sealed. The slightest leak allowing air to enter ruins the food. Also, some bacteria have a nasty trick of occurring in two forms, a vegetative form easily killed by heat and a spore form which is heat-resistant. The worst organism of this type is *Clostridium botulinus*. It occurs in soil and grows only in the absence of oxygen. Food in a can or bottle which has not been adequately heated will provide this organism with a chance to multiply. It produces the most deadly poison known and the poisoning is difficult to treat, though an antitoxin does exist. Therefore meat, fish and vegetables must be heated either to 240° F (115° C) or kept at the boiling point of water long enough to kill spores.

The simple canner, which is merely a large container with boiling water in the bottom, is suitable for canning acid fruits. These fruits do not become contaminated with botulinus because the acid juice prevents growth of the organism. Non-acid vegetables, especially those that have been in contact with soil (such as carrots or beets) are more safely preserved in a pressure cooker. This instrument should be one of the valued tools of the Whole-earther. It makes possible the safe preserving of practically anything. An open canner is worthless at high altitudes because the higher you go, the lower the temperature at which water boils. The pressure cooker can compensate for altitude. He who lives above the 2,000-foot level should add 1 pound of pressure for each additional 2,000 feet.

Anyone using a pressure cooker should understand a few simple principles. Food may need precooking. Meat, for instance, should be precooked until red color changes to brown but flour should not be used. Vegetables are precooked in a minimum of water, which can be used to cover the product in the jar. Fish, cut into suitable-size pieces, can be lightly fried. A pressure cooker uses the pressure of *hot steam*, not *hot air*. For this reason the jars must be placed in the cooker filled to within an inch or half inch of the top, and steam allowed to flow from the petcock for 7 to 10 minutes. In this process all the air in the jars is replaced by steam. If air is left in the jars it will force the liquid out when the pressure is applied and partly empty jars will be the result. Because of this long exhaust period, at least 2 quarts of water must be placed in the bottom of a 7-quart cooker. The jars should be placed in a basket, not on the floor of the cooker. The two-piece metal cap can be screwed down firmly before the jar is placed in the cooker. Air can escape because of expansion of the cap.

A pressure cooker is a potential bomb and should therefore be handled with respect. *Never* allow pressure to rise until you are sure the lid is on properly. *Never* use a cooker with a crack in it or one that has become distorted because allowed to boil dry. *Never* try to remove the lid until pressure is down and don't try to hasten the process by releasing the petcock or liquid will boil out of the jars. Watch the pressure gauge to make sure it remains steady. Timing begins from the moment when the desired pressure is reached. At the end of the process, remove cooker from the source of heat. After the pressure has sunk to zero on the gauge, cautiously open the cooker, allowing steam to escape on the side away from you. Bare-armed operators commonly scald themselves at this point so wear a jacket. Remove the jars from the cooker and allow them to cool. Tap lids when cool. A high-pitched ping denotes success, a dull thud failure. If a jar has not sealed it won't keep so its contents should be used at once.

For processing times and pressures, see *The Modern Guide to Pressure Cooking and Canning* (National Pressure Cooker Co., Eau Claire, Wis. 54701). In general, acid fruits like apples, pears, plums and peaches are precooked in medium syrup (1 cup sugar, 2 cups water) and processed for 10 minutes at 5 pounds of pressure. Vegetables in quart jars range from 10 minutes at 5 pounds for tomatoes to 90 minutes at 10 pounds for baked beans and winter squash. For all meats placed without bones in quart jars, 90 minutes at 10 pounds is suggested, 65 minutes at 10 pounds in pint jars. Fish in pint jars need 80 minutes at 10 pounds but for salmon 100 minutes is recommended. Shad should also be cooked for 100 minutes. In this way the innumerable small bones that are the curse of this fish can be made edible.

The pressure cooker can, of course, also be used for cooking meals and is valuable for the preparation of hard-to-cook items like tongue, pig's feet, dry beans and beets. The large 7-quart instruments suitable for canning are not good for cooking. For this a smaller cooker is desirable. Such cookers (Presto, 4- or 6-quart capacity) have a rocking-type pressure regulator that automatically releases steam and keeps pressure at 15 pounds. Cooking by pressure has three great advantages: steam does not leach out soluble substances from the food as does boiling water, oxygen is excluded and the high temperature ensures rapid cooking. An ingenious cook can, by suitable arrangement of pans and colanders, cook several items in the pressure cooker at the same time. The Whole-earther getting ready for the Fossil Fuel Failure should stash away one or both of these instruments (the 7-quart canner and the 4- or 6-quart cooker). They are one of industry's finer contributions to life.

3. Pickling, jam making

Pickling is one more device that emerged in the course of man's long struggle to preserve his food. The technique depends for its success on the fact that vinegar is acidic and inhibits the growth of most organisms that cause food spoilage. The complete Whole-earther may wish to make his own vinegar (see section 7). If he buys it he should make sure that it contains 4–6% acetic acid. This is generally marked on the label. Cider vinegar is generally considered best. In many pickles salt as well as vinegar is used as the preservative, or salt may be used by itself in the form of brine. Some meats and even vegetables such as green beans can be preserved by packing them in dry salt. The salt draws liquid out of the meat or vegetable, forming a brine sufficiently concentrated to preserve the food. Such food must be soaked free of salt before eating.

A typical vinegar pickle is the pickled beet made by covering cooked young beets with a mixture containing 2 cups each of sugar, water and strong vinegar with a teaspoon each of cloves and allspice and a tablespoon of cinnamon. The combination is boiled for 10 minutes, then sealed in jars. Onion pickles involve both salt and water. The small onions are soaked in a brine (1½ pounds of salt to a gallon of water) for 5 weeks, drained, rinsed and preserved in a boiling mix of equal parts of vinegar and sugar. The mix is drained off, reheated, poured back over the onions on 3 consecutive days. They are then covered with the mix, a teaspoon of pickling spices added and the jars sealed.

Dill pickles are made by soaking small cucumbers in a brine (1½ cups of salt to 4 quarts of water) overnight. A second brine is made with 10 quarts water, 1 quart vinegar, 2 cups salt, which is boiled for 10 minutes. Cucumbers are packed in jars with bunches of dill and covered with the cold canning brine. They will ferment for 3–4 days and become clear in the process. The same process can be used on a larger scale, with a crock or barrel.

Jams and jellies. This form of preserved food depends on the fact that sugar in high concentrations inhibits the growth of most molds and bacteria. Honey keeps for this reason. The jam maker simply mixes fruit with an equal weight of sugar, boils the mixture until concentrated, pours the product into heated, sterilized jars and seals the surface with a layer of hot wax. Fruits like blackberries are better precooked and strained to remove seeds, then mixed with sugar. The seeds are indigestible and add nothing of value to the product. This method gives a jelly rather than a jam but the gel forms only if fruits contain enough pectin. Pectin is a gelatinous substance present in fruits in variable amounts. Pectin will not gelatinize unless acid is present. It is just as well, for this reason, to add lemon juice to the mix if a firm gel is required. It is also possible to buy pectin preparations that can be added to fruits such as apricots or peaches which contain very little.

4. Freezing

The best method of storing food is by freezing, though the true Whole-earther who wishes to avoid depending on utility companies will probably avoid the method. A freezer is utterly dependent on a reliable source of power and if the power fails the food rots. The method depends on the fact that neither molds nor bacteria will multiply at temperatures around 0° F so frozen food needs no preservative. The enzymes in fruits and vegetables continue to operate at these low temperatures, for which reason, to avoid the development of bad flavors, the operator heats these materials for about 2 minutes in boiling water, cools them and packs in suitable containers. Meat and fish do not need this pretreatment, but meat, especially game, must be allowed to age preferably for 3 days before being cut up and frozen. Fish is best fileted before freezing to save space. Meat should have the bones removed for the same reason. The bones can be boiled to form a concentrated stock, which can also be frozen.

5. Butchering

A large animal (steer, sheep or deer) must be skinned quickly. Delay makes the job more difficult. Start the job on the floor. Work from the hind legs but don't sever the tendon of Achilles or you will have nothing on which to hang the corpse. As soon as the hind legs are skinned, put a wooden or metal support through the gap between tendon and bone and hoist the carcass aloft. To lift a steer you will need a pulley. Strip off the skin from back and belly. Be careful not to cut through belly muscles. Complete skinning. Carefully open the abdomen (a burst stomach stinks and makes a mess). Separate the liver from the guts. Leave kidneys in the carcass. Saw down the breastbone. Remove heart and lungs. Hold the carcass open with wooden skewers and cover with cheesecloth to keep off flies. Let it hang in a cool place for 3 days.

With carcass still hanging, saw down the center of the backbone and cut across the backbone on the level of the last rib. This reduces the beast to manageable proportions. From there on, follow the anatomy to get the standard cuts (fig. 12), from which bones can then be dissected. Wrap the meat in freezer paper and label it before putting in the freezer.

6. Natural dry foods

There are plenty of foods that need neither canner, freezer, vinegar nor sugar to preserve them. They keep because they are naturally dry. The true Whole-earther who really wants to be

92 FOOD

free of trouble need only lay in a stock of these dry foods and can then devote his whole time to meditation or any other activity that takes his fancy. Examples are brown rice $12.20 per 100 pounds, red kidney beans 17¢ per pound, split peas 17¢, red Montana wheat $9 per 100 pounds, soybeans $8 per 100 pounds. Check feed dealers, grain dealers, seed dealers. The keep-a-year's-supply Mormons offer a year's food supply for around $190 (Perma-Pak, 40 E. 2430 South, Salt Lake City, Utah 84115). These prices, of course, are subject to inflation.

The four-component diet suggested by Esther Dickey (*Passport*

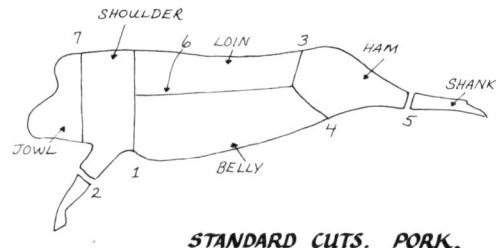

STANDARD CUTS. PORK.

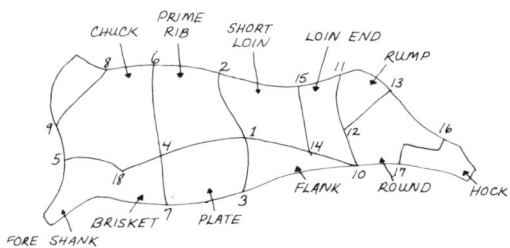

STANDARD CUTS. BEEF.

BUTCHERING PORK AND BEEF

FIGURE 12

to *Survival,* Bookcraft Publishers, 1848 W. 2300 South, Salt Lake City, Utah) contains wheat, powdered milk, honey and salt. A month's food supply for one adult is 27 pounds wheat, 5 pounds powdered skim milk, 3 pounds honey, up to a pound of salt. Fresh greens can be obtained by sprouting the wheat. Yeast, a fine source of B vitamins, can be grown in a mixture of wheat flour and honey.

The true Whole-earther may be reluctant to depend on a product of the Great Society such as powdered skim milk. Substitution of soybeans for the milk will result in a fairly well balanced diet. The soybean can be sprouted, in which case it rapidly develops a rather high level of vitamin C. It is high in protein and fat and contains adequate levels of calcium. Soybeans are the basis of MPF (Multi-Purpose Food) developed by Dr. Henry Borsook of the California Institute of Technology. It is made of soy grits (86%) with onion powder, protein derivative and food yeasts, plus calcium pyrophosphate and vitamins. Two ounces of MPF is equivalent to an ordinary meal. It is distributed by Meals for Millions Foundation, 115 W. Seventh St., Los Angeles, Calif. Any serious hermit intending to meditate for a year or two in a cave would do well to lay in a supply of MPF. He will certainly stay healthier on MPF than did Tibet's great hermit Milarepa on an exclusive diet of nettles.

7. Fermenting

Fermentation is the process which occurs when a food substance is acted on by yeasts, molds or bacteria and changed in some way. Generally speaking, fermentation is exactly what the food storer tries to prevent, for fermentation is only a step away from putrefaction that renders the food uneatable. There are, however, certain fermentations that men have used ever since Neolithic times to create special foods. Wine making, brewing, vinegar making, cheese making and bread making all involve fermentation and the true Whole-earther should understand the processes involved.

Wine making. Wine, which has been making glad the heart of man ever since the days of Noah, results from the action of yeast on fruit sugar. The product is ethyl alcohol, the effects of which are too well known to need description. The wine maker can be reasonably successful if he/she remembers certain facts:

(1) There are good yeasts and bad yeasts. The bad ones produce, in addition to ethyl alcohol, substances like fusel oil that taint the wine and make it undrinkable. To get rid of bad yeasts, add good yeasts in such amounts that they overwhelm the bad. Baker's yeast, brewer's yeast and wine yeast will all ferment fruit juice, but wine yeasts selected for their special qualities are best. These yeasts can be obtained as packets of dry powder from suppliers such as Wine Art of America, 4324 Geary Blvd., San Francisco, Calif. 94118.

(2) A packet of wine yeast does not have much of a chance of starting a good fermentation if you drop it at once into several gallons of fruit juice. Dry yeast consists of a mixture of living and dead cells and the living cells multiply best in a small volume of liquid. Prepare your "yeast starter" by sterilizing a bottle with 10% sulfite solution and rinse with boiled water. To half a pint of the material you wish to ferment, add a tablespoon of sugar, a teaspoon of citric acid and a pinch of potassium phosphate or ammonium phosphate. Sterilize this solution by boiling, cool, add the yeast powder, close bottle with a plug of cotton. Keep in a warm place (75 to 80° F) until fermentation begins.

(3) If the must you are going to ferment is obtained by boiling some fruit with water and sugar, this solution, when cool, will be reasonably sterile and the yeast starter can be added to it directly. If you are dealing with large volumes of crushed fruits (grapes, for instance), it is best to partly sterilize the must with sulfite. Sulfite is short for sodium metabisulfite. It is convenient to obtain it in the form of Campden tablets. One or two tablets per gallon of must or 5–10 milliliters of a 10% stock solution will introduce enough sulfur dioxide into the must to kill the undesirable organisms. The sulfite should be thoroughly stirred in and the must left for 24 hours. Then the yeast starter can be added.

(4) Cleanliness of glassware and barrels plays an important

role in determining whether the wine maker ends up with a drinkable product. Glassware can be cleansed and partly sterilized by rinsing with a 10% solution of sulfite and then with hot water. Barrels can be sterilized by burning a sulfur candle inside the *wet* barrel or filling with water to which sulfite has been added. Barrels should always be stored full of water to which sulfite has been added.

(5) Air is the enemy of wine and the friend of vinegar. Therefore he who wants wine rather than vinegar should keep the air out of the fermentation. Fill the vessel as full as you can without having the must froth over. Stopper with a water lock which lets the carbon dioxide out but prevents air from entering. If you make 50 gallons at a time a barrel is a good container but should not be filled completely until the first vigorous fermentation has ended. For smaller amounts 5- or 10-gallon water bottles are sufficient. Always have enough extra must fermenting in a bottle to top up your barrel before racking.

(6) The amount of alcohol a yeast can produce depends on the amount of sugar in the must. A grape must is not considered worth fermenting until it has at least 18% sugar. A hydrometer is used to determine the sugar content of must. It floats in the solution and the higher it floats, the more sugar the solution contains. What the hydrometer reads is specific gravity which, with the help of a table, can be converted to percent sugar. A reading of 1.090 equals a solution containing 24% sugar and one of 1.075 equals 20%. These two solutions after fermentation will give 14.4% and 12% alcohol respectively. If the must contains too little sugar this substance must be added either as sugar or honey. Small amounts of sugar (2 ounces per gallon must) can be added *in solution* as fermentation slows to boost the alcohol level of the wine.

(7) Completion of fermentation leaves the wine with a thick layer of sludge in the bottom of the container. This yellowish sludge consists chiefly of dead yeast cells. Dead yeast cells have a way of dissolving themselves (autolysis) and the products of their dissolution often taint the wine. For this reason the wine has to be

siphoned into another clean and sterilized vessel and the layer of yeast cells left behind, a process known as racking (fig. 13). If you are operating on the 50-gallon-barrel scale it is best to rack into 5-gallon bottles and pour these into the second barrel, previously sterilized with sulfite. Some wine will be lost in racking so the barrel must be topped up from one of the carboys of juice separately fermented for this purpose. Fermentation may begin again after the first racking and a second racking may be required.

(8) As fermentation proceeds specific gravity falls. At a grav-

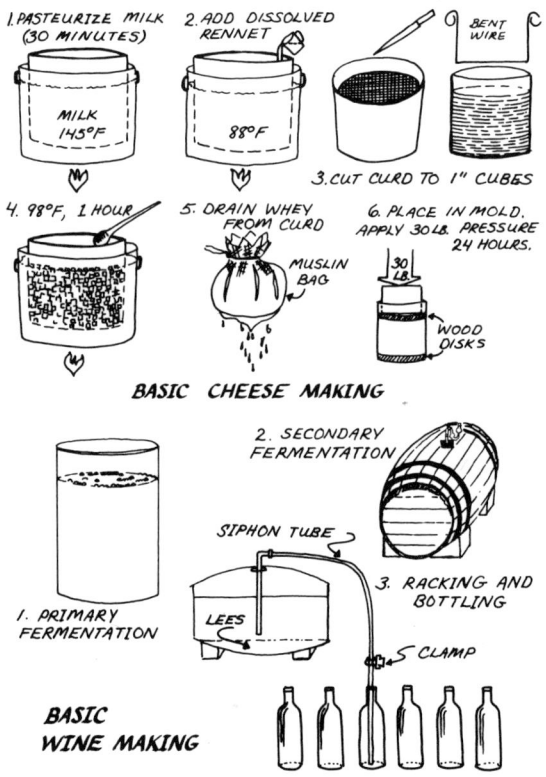

FIGURE 13

ity of about 20 a sweet wine is produced (*gravity* is calculated from *specific gravity* by deducting 1 and ignoring the decimal point; gravity 20 = specific gravity 1.020). If a sweet wine is desired further fermentation can be prevented by adding sulfite (one Campden tablet per gallon) to the wine after racking. Subsequent rackings at intervals of 2–3 weeks will prevent the yeast from increasing. A light sweet wine of the *Liebfraumilch* type can be produced in this way.

(9) The end of fermentation does not mean that the wine is made. It needs a period in which to mature and this maturation is best accomplished in a wooden cask. For the small-scale operator a cask of 9–16-gallon capacity is satisfactory. The cask should be completely filled, closed tightly and left in a quiet dark place at a temperature of 55–60° F.

(10) Bottling confronts the wine maker with his most difficult problem. The question is when to bottle and no single answer can be given because wines differ enormously in the time they require in the cask. A sweet white wine of the *Liebfraumilch* type will be ready for bottling after 6–12 months, but a full-bodied claret may take 20 years to reach its peak quality. A sample of wine drawn from the barrel into a bottle and stored, plugged with cotton, for several days at 75–85° F should not re-ferment, nor should it form a deposit. Bottling before fermentation is complete is a sure way to end up with blown corks. Bottling itself is a simple operation. The bottles must be washed with detergent and hot water, carefully rinsed and left upside down to dry. They can be filled with a siphon that can be slowly lowered into the barrel. Care must be taken not to disturb sediment. Each bottle is filled to within ½ inch of the point occupied by the bottom of the cork. Corks can be softened and sterilized by boiling in water. Suppliers of wine-making equipment can provide the device necessary for pounding in the cork.

(11) Many wines after maturation in wood still remain cloudy. The cloudiness does not affect the flavor of the wine but it does deprive it of sparkle. The amateur wine maker has to decide just how much effort he is willing to make for the sake of appearance.

He can try to clear his wine with fining agents (white of egg, gelatin) or he can filter it. Filtration tends to overaerate wine unless precautions are taken. Bottling will aerate wine to some extent in any case and all wines should be left in the bottle lying on the side for 3–6 months to recover from what is called bottling sickness.

I have found that some wines, particularly apple wine or cider, taste much better if not matured in wood. Apple wine is best made by crushing the apples in a cider press, fermenting under a trap in glass bottles and racking once. This product does not need to age. It is at its best about 6 months after preparation.

Brewing. The process of brewing resembles that of wine making except that grain is used as a source of sugar instead of fruit juice. Beer of a sort can be made from many kinds of sprouted grain but malting barley is best. The amateur brewer can create a hearty, heavy brew by the following procedure:

(1) Bring 7 quarts of water to 150° F. (If the brewer has the misfortune to depend on a municipal water supply he had better boil the water first to get rid of added chlorine.)

(2) Add 2 pounds crystal malt, 2 pounds patent black malt. Cover the vessel with a sheet of polyethylene and wrap it in a blanket to conserve warmth (a polyethylene pail makes a good container).

(3) Switch on the heater, maintain the mash at 145–150° F for 8 hours. Strain this mash into a boiler and add 2 level teaspoons of salt and 2 ounces of dried hops. Bring to the boil. Simmer gently for 40 minutes. Add one more ounce of hops and boil for a further 5 minutes.

(4) Prepare the fermenting vessel and place within it 3 pounds white sugar, 1 pound black treacle, ½ ounce citric acid. Strain the mash onto it through fine muslin. Stir, making sure all sugar is dissolved. Make up to 4 gallons with boiling water.

(5) Cover with a plastic sheet. Cool to 65–70° F. Add brewer's yeast and yeast starter. Allow to ferment 6–8 days in cool place (60° F or lower).

(6) Take readings after 5 days. When the hydrometer regis-

ters 1.005, bottle. If you don't have a hydrometer, leave the beer until it goes flat. You can put back the fizz by adding a little more sugar. Don't overdo it or bottles may blow up.

This operation results in a product called Bravery's Super Stout, a far cry from the washy stuff foisted on the public by American brewers. For more recipes, see *Home Brewing Without Failures* (H. E. Bravery, Arc Books, 219 Park Ave. South, New York, N.Y. 10003), or *Last Whole Earth Catalog*. The brewing supplies (according to a correspondent of the *Last Whole Earth Catalog*, p. 202) can be obtained from Wine-Art Sales Ltd., 1108 Lonsdale, North Vancouver, B.C., Canada.

Vinegar making. For the wine maker the main problem is how *not* to make vinegar. The fermentation that turns wine or cider to vinegar is brought about only too easily by an organism, called *Mycoderma aceti,* that oxidizes alcohol to acetic acid. The "mother of vinegar" is best obtained from a good, non-sterilized vinegar which should be mixed with the wine as a starter. The wine or cider can be placed in a barrel (which should on no account be later used for wine) and kept at about 76° F. The acetic acid forms slowly by this process. It takes 12 to 16 weeks. But the vinegar so formed will have the bouquet of the original wine or cider and be far superior to the synthetic product, which is merely a solution of acetic acid. The vinegar process can be hastened by letting the wine or cider trickle slowly over a bed of charcoal or hardwood shavings. Only wine containing less than 14% alcohol should be used for vinegar making.

Bread making. When flour and water are mixed and the dough is baked the result is a stodgy mass hard to swallow by anyone not absolutely ravenous. This is unleavened bread. The "leaven" which lightens this lump is a yeast that has the power to ferment some of the starch in the dough to alcohol and carbon dioxide. Bread is made by mixing flour, yeast and water, warming until the dough rises due to formation of bubbles of carbon dioxide, and baking. The baking process drives off the alcohol. The bubbles of carbon dioxide are trapped by the elastic gluten, which is hardened by heat. The result is a light-textured substance, easily chewed and digested, a far cry from the soggy, unleavened product that was all primitive man had until he discovered the yeast fermentation.

A good, nourishing bread can be made with 7 cups of whole wheat flour (or 4 cups whole wheat and 3 cups white for less rugged consistency). A mix is made with ¾ cup milk, 3 tablespoons sugar, 4 teaspoons salt, ⅓ cup oil, ⅓ cup molasses, 2 cakes of baker's yeast and 1½ cups warm water. Stir in 2 cups whole wheat flour and 2 cups white flour. Beat until smooth. Add enough of the remaining flour to make a soft dough and knead thoroughly on a lightly floured board. Place in greased bowl and allow to rise in warm place (about 90° F) until doubled in volume. The dough must then be punched down, divided in portions and placed in greased bread pans. Allow to rise again until almost double, then bake. This dough will make two loaves 9 × 5 × 3 inches. Bake at 350° F for 35 minutes. Addition of ½ cup of soy flour will improve the protein content.

Baking bread can be done without gas or electricity in a brick oven heated with wood. However, the brick oven or old-fashioned kitchen range is a heavy fuel consumer and the true Whole-earther, eager to conserve fuel, may prefer to make his bread in a manner that takes less heat. This can be done by using a frying pan or skillet and making pan bread, tortillas or chapattis. A good tortilla mix contains 1 cup wheat flour, ½ cup cornmeal, ¼ teaspoon salt, 1 egg, 1½ cups cold water. Beat into a batter, drop on a hot, ungreased griddle. This will make 12 tortillas, unleavened but good, especially with a filling of boiled beans.

Florida hush puppies (so called because Florida hunters would toss fragments of this tasty bread to their dogs to keep them quiet) require no oven. Just beat together 3 cups cornmeal, 2 teaspoons baking powder, 1½ teaspoons salt, 1½ cups milk, ½ cup water. Add an egg if you have one and a finely chopped onion if you like onions, mold into little cakes, fry in about an inch of fat until brown.

If all else fails, the hungry man can make a very good bread substitute in the form of dumplings. Mix 1½ cups wheat flour, ¾ cup soy flour, 2 teaspoons baking powder, ¾ teaspoon salt, 3 teaspoons oil and ¾ cup milk. Drop carefully on the *surface* of boiling water, or soup if you have it. Cook for 10 minutes with pot uncovered and 10 more with it covered. Addition of an egg improves the product.

Breads can also be made by steaming instead of baking. A good steamed bread can be made by mixing 1 cup rye flour, 1 cup cornmeal, 1 cup whole wheat flour, 2 teaspoons soda, 1 teaspoon salt. Add ¾ cup molasses and 2 cups sour milk. Put in two 1-pound coffee cans, cover with wax paper and steam for 3 hours.

Cheese making. Cheese is one of mankind's oldest foods. The fermentation which produces it takes place naturally whenever milk is allowed to stand in a warm place. Bacteria in the milk convert milk sugar (lactose) to acid (lactic acid). This acid causes the protein in the milk (casein) to curdle. The product of this curdling, if the bacteria used are *Bacillus bulgaricus* or *B. acidophilus,* is yogurt, a food in itself. There is no scientific evidence to support Eli Metchnikov's view that the eating of yogurt prolongs life. It is a fact, however, that some people have a hard time digesting milk sugar and yogurt is free of this material.

If you don't want to be scientific about cheese making and are satisfied with peasant techniques, just put a few gallons of fresh milk (cow or goat) in a clean crock, cover to keep out flies, stand in a warm place. When the milk has soured, pour into a conical bag of cheesecloth and let it drain. The whey (liquid part) makes good pig food or you can boil it down to make whey cheese. It contains nearly all the minerals and a lot of other goodies from the milk, but making whey cheese is rather costly in fuel.

The curds left in the bag are the simplest form of cheese (cottage cheese) and can be eaten as such. If you put the curds in a press and squeeze out the remaining whey, adding salt and any flavoring agents you fancy, you get a cheese which will keep for several weeks if refrigerated.

If you wish to be more scientific about cheese making a few special items of equipment will be needed:

(1) A utensil large enough to hold 5 gallons of milk. It can be made from a 5-gallon oil can. Slit the can lengthwise. Cut from each end of the slit to the corners making two flaps. Roll the flaps to the sides.

(2) A large utensil to hold the 5-gallon can and act like the outer member of a double boiler.

(3) A cheese cutter, which can be made from a 3-foot piece of baling wire. Bend both ends of the wire at right angles so that the bottom of the wire is the width of the 5-gallon can. The two ends should extend above the utensil to give hand holds.

(4) A long-bladed knife.

(5) Cheese hoop and two followers. The hoop can be made from a #10 can with the bottom removed. The followers are simply wooden discs which fit inside the can.

(6) A press. You can use a car jack (hydraulic type) or a simple lever device.

(7) Cheesecloth, measuring spoons and a thermometer with a scale as low as 60° F.

In addition you will need 5 gallons of milk, a cheese rennet tablet, salt and a starter. The starter can be a carton of churned cultured buttermilk, *not* pasteurized. Or you can buy starter from a creamery.

Your 5 gallons of milk will make a 5-pound Cheddar-type cheese if you follow these steps:

(1) Wash and scald the equipment.

(2) Put the milk in the 5-gallon can and place in the outer boiler (fig. 13). Heat the water in the outer boiler until the milk in the can reaches a temperature of 140–145° F. Hold it at this temperature for 30 minutes. This pasteurizes the milk.

(3) Put cold water in the outer boiler. Bring down the temperature of the milk to 88° F. Add ½ cup of starter. Hold the milk at 88° F for 20 minutes.

(4) Add ½ rennet tablet in ¼ cup cold water. Cover and let stand at 88° F at least 30 minutes.

(5) Test the curd. It should be firm enough to break away cleanly from the sides of the container or break over the finger when it is inserted into the curd. Cut the curd into half-inch cubes, using the baling wire and long-bladed knife. Cut curd lengthwise and crosswise with the knife, then make a series of cuts with the baling wire parallel with the bottom. Let stand after cutting for 3 minutes, then stir gently and wait for 10 minutes before heating.

(6) Slowly heat water in outer boiler to bring temperature of cheese to 100° F. The heating should take about 30 minutes. Keep curd in slow motion with a spoon throughout the heating period to prevent cubes of curd from sticking together. Hold curd at 100° F for an hour, stirring occasionally. The curd should be firm but fall apart with light shaking when it is squeezed in the hand. It makes a squeaky sound when chewed.

(7) Drain off the whey and pack the curd in one end of the can. Tip the can so that remaining whey will drain into the empty end. Remove whey. Cut curd with the knife into 1-inch cubes and mix in 4 level tablespoons of salt. Pack down and let stand for a few minutes, then cut again into 1-inch cubes.

(8) Place a wooden follower in the bottom of the cheese hoop. Line the hoop with cheesecloth and pour in the curd. Fold cheesecloth over the curd and place second wooden follower on top of the cheese hoop. Put in cheese press and apply weight equal to 60 pounds pressure for 30 minutes.

(9) Remove cheese from press and dress in cheesecloth. To do this, cut two circles for the top and bottom of the cheese. Cut a strip long enough to go round the cheese and wide enough to meet at the center of the top and bottom. Dampen the cheesecloth and apply to the cheese, then return cheese to cheese hoop. Reapply pressure. If the surface of the cheese was not smooth and free of breaks, increase pressure to 100 pounds and press for 24 hours.

Cheese, like wine, needs to age, so the cheese after removal from the press should be placed on a clean, dry shelf and stored at a temperature of 50–63° F. Turn the cheese over every day and wipe it dry with a cloth. In 3–7 days, when the outside surface is dry, coat the cheese by dipping it in hot wax. Replace cheese on the shelf in the cool room. Under these conditions the cheese will be ready to eat in about 4 weeks. Once cut it should be stored in the refrigerator. Salad oil rubbed on the cut surface tends to prevent drying.

Cheeses like Roquefort or Camembert have special molds growing in them that impart a characteristic flavor. The molds can be obtained from these cheeses themselves or from the American Type Culture Collection, 12301 Parklawn Dr., Rockville, Md. 20852. The mold is mixed in with the curd before the latter is placed in the cheese hoop. Cheese can be made from goats' milk or sheep's milk as well as cows' milk. Every Basque shepherd who takes his flock up to the high pastures in summer also takes his own wooden cheese press, from which he makes cheeses of ewes' milk with his own personal imprint.

Cheese can also be made from soy milk. Soybeans are washed and soaked in cold water overnight, using 2 quarts of water for a pound of beans. The beans, separated from the water, are ground in a food chopper as finely as possible. The water is replaced and the milky mass poured through cheesecloth. The pulp that fails to pass the filter can be used in cooking. The soy milk is quite a good milk substitute, comparing favorably with cows' milk as far as protein, iron and B vitamins are concerned. It can be made almost exactly like cows' milk by the addition of glucose and calcium hydrogen phosphate. Soy milk is suitable for babies that are allergic to cows' milk. It is much cheaper and you don't have to keep a cow. Soyalac is a spray-dried soy milk powder.

To make soy cheese from soy milk, add 1½ parts vinegar to 100 parts milk, or sour the milk with a culture of *Bacillus acidophilus*. A solid precipitate forms, which is soy curd. It is allowed to sink to the bottom of the container and the liquid above it is discarded. The curd is transferred to cheesecloth in a colander to drain off remaining water. The curd can be molded and pressed to form small cheeses. Such cheeses are further fermented in the East by the introduction of special molds (obtainable from the

Northern Utilization Research Branch, Peoria, Ill.). Similar strains of mold are used in the formation of soy sauce. Soybean curd contains 7 to 9% of highly digestible protein and has been aptly described by the Chinese as "meat without bones." Vegetarians should avail themselves of this excellent cheap protein source, particularly when it comes to food for pregnant women and for children, whose protein needs are urgent and often unsatisfied by vegetarian diets.

Butter making. Butter making does not necessarily involve fermentation. A "sweet" butter can be made simply by separating cream from milk, agitating it in a suitable churn (a gallon churn made of glass with wooden paddles will churn 2 quarts of cream), draining off the buttermilk (which will be sweet because unfermented) and washing and working the butter. About 1 tablespoon of salt per pound should be worked into the butter, which can then be wrapped in parchment paper and stored.

It is somewhat easier to make butter if the cream is "ripened," or soured. You can use for souring a sample of buttermilk that has not been pasteurized or simply add a cup of freshly soured milk to the cream. It is desirable first to pasteurize the cream, as one pasteurizes milk for cheese making. Cream will sour at 65–70° F in about 24 hours. Churn the cream at a temperature of 58–66° F in winter, 52–60° F in summer, using ice if necessary to bring it down to these temperatures. Butter will form in about 30 minutes. Strain off the sour buttermilk, wash the butter, add salt and store in refrigerator.

8. Cooking

Man is the only creature in the biosphere that cooks meals. The practice goes back beyond the Neolithic revolution to the time when man first discovered how to make fire. Fire, the gift of Prometheus, was above all a warming agent. How man first learned to cook with fire is a matter for speculation. Perhaps, as Charles Lamb suggests in his *Dissertation on a Roast Pig*, the discovery originated when someone's grass hut caught fire and a pig or some other item of livestock perished in the flames. The man, feeling reluctant to waste good food, chewed on the partly burned corpse, discovered it was easier to eat, tasted better, was indeed a totally different substance from the raw and bleeding meat that could only with difficulty be torn apart and chewed.

What was true of meat also proved true of vegetables. A host of roots and grains that were indigestible raw changed, when cooked, into quite different substances, with different consistency and a different flavor. The cook was the first chemist and cookery gave man his earliest experience of manipulating matter to suit himself. He used, without knowing it, the properties of infrared radiation or elevated temperature which will break down many large molecules (like the protein of meat) into smaller ones. This makes the food easier to chew and easier to digest.

Enthusiasts of the back-to-nature school may argue that cooked food is spoiled food, that it is perfectly possible for man, if he selects his food carefully, to live without cooking. Nuts, various fruits and vegetables, even beans and wheat if they are sprouted, can be eaten raw. So can eggs and fish, and meat if it is sliced thin enough or dried in the sun and pounded with berries into pemmican. This is quite true. It is also true that cooking may damage food either by leaching out minerals and vitamins or by destroying them with heat. Furthermore, an inordinate amount of time may be spent cooking food, time which might be devoted to other activities. For one who does not enjoy cooking, the daily chore over the stove may be a pain indeed.

So let us be clear about this. Man can live without ever cooking his food. A good meal of sprouted soybeans, sunflower seeds, milk, honey, raw carrots and some fruit will give the non-cook enough nourishment without his having to go near a stove. But it will also place beyond his reach several goodies that are almost uneatable raw, potatoes, for instance. And the non-cook should also remember that cooking, as well as changing the state and consistency of food, sterilizes it. The organisms causing dysentery (both amebic and bacillary), typhoid fever, cholera, various forms of food poisoning and trichinosis are all destroyed by heat. So is the deadly toxin of botulinus that sometimes forms in home-canned products. No one will die of botulinus poisoning who brings his home-canned beans or beets to the boil before eating them.

Damage to food by cooking can be kept to a minimum if certain facts are borne in mind. Proteins, which make up the bulk of meat, fish and such vegetables as soybeans, are denatured by cooking, which means they are irreversibly altered in structure. This is obvious to anyone who has ever fried an egg. The liquid, colorless albumen changes to a solid white substance and the yolk also becomes solid. Meat becomes more tender as it is cooked. An old chicken needs a longer stewing than a young one because its protein is more strongly cross-linked than the protein of a young bird. Pig's trotters or ox tongue can be rendered edible only by hours of cooking at ordinary temperatures or drastic heat treatment in a pressure cooker. Dry beans also need long cooking.

This prolonged heat treatment, far from decreasing the food value of the protein, actually increases it, by reducing the big protein molecules to smaller fragments which can be more easily digested. This applies also to starchy vegetables like potatoes, sweet potatoes, yams. The big starch molecules are partly broken down and rendered more digestible and the cellulose cell walls of the vegetables are also partly disintegrated.

There are some proteins, however, that are not improved by long cooking. Seafood such as mussels and fish reach a point at which further application of heat makes the protein tough. The same applies to eggs. An omelette rapidly acquires the consistency of leather if cooking is prolonged beyond the critical point.

A question of special interest to the Whole-earther is how to cook with a minimum of fuel. In rich, overdeveloped countries such as the United States, cooks rarely even pause to consider this question. They use electricity or gas and assume that the supplies of both commodities are inexhaustible. A very different situation occurs in poor countries such as India. There is neither gas nor electricity in the village. The villager cannot even go into the forest for a load of dead sticks. There is no forest. It was cleared long ago to make way for fields. So the cook has to use dry cow dung, which deprives the fields of needed manure and makes a poor fuel anyway.

Solar cookers and other cheap methods of heating food will be described later in this section. Here it need only be said that fuel economy will influence the Whole-earther's choice of cooking methods. Roasting and baking are both very heat-consuming processes and, unless the stove is being used to heat the kitchen, so that the oven is hot anyway, they should be avoided. Gas ovens and electric ovens both guzzle huge amounts of fuel. The true Whole-earther will avoid using either.

In communes where the key word is simplicity much time can be saved if relatively few recipes are used. I personally live on much the same food all the year round but find infinite variety in the salads and vegetables, not to mention main dishes. The variety is found not by varying the ingredients but by eating consciously and absorbing impressions as well as food. Eating, like anything else, can be done well or badly. The eater can shovel in his food and be quite unaware of the process, reading the paper, listening to the radio, jabbering to his wife. Or he can intentionally savor the good substances provided by the biosphere, remembering that his life depends on them. He can distinguish the four tastes, sweet, sour, salt and bitter, the innumerable categories of aromas, the various textures of his food. He can remember where the food came from (and if he is a true Whole-earther he knows this well for he grew the stuff himself). He can reflect on the chemical transformations which turn sunlight into food and turn food into him. If he absorbs it in this spirit he will never get bored with his food.

A fine way to start the day and fuel the machine is with a pancake or waffle prepared according to the following recipe. It is taken from *Let's Cook It Right* (Adelle Davis, Harcourt Brace & Co., New York, 1962), a real Whole-earther's cookbook. Adelle Davis scorns baking soda. Use yeast, she says, even for waffles and pancakes. The same mix will do for both. For waffles:

(1) Stir one package or cake of crumbled baker's yeast into 2 cups of warm water, warm milk, buttermilk or yogurt.

(2) Add 2 tablespoons molasses or sugar, 3 egg yolks, 1 teaspoon salt, ⅓ cup shortening or oil, 1 cup wheat germ.

(3) Sift in ¾ cup whole wheat flour, ½ cup soy flour, ⅓ cup powdered milk.

(4) Stir well, let rise in a warm place for 2 hours or longer, stirring down each time batter has doubled in bulk.

(5) Just before baking, beat stiff and fold in 3 egg whites.

Any batter left over can be thinned with milk and used for pancakes. There is no need to heat up the stove to cook these daily. A whole week's supply can be prepared and stored in a cool place, wrapped in waxed paper. One waffle will provide a working man with enough calories for the morning's work. Such waffles or pancakes are bread substitutes that can be made without using the oven. Spread on them a layer of peanut butter, a layer of homemade blackberry jam or a layer of apple butter flavored with fresh mint. That's a gift for the gods. This breakfast will give protein, fat and carbohydrate, all of the needed vitamins will be added with a glass of milk, or, if you don't keep livestock, a glass of vegetable juice or some grated raw carrots.

For the non-meat eater the bean (soy, lima, navy, kidney, etc.) is the surest defense against protein deficiency. Soybeans can best be prepared by soaking and freezing. The procedure reduces cooking time by about 2 hours. The following recipe may not be to everyone's taste but will definitely provide protein in plenty for the vegetarian. Take 1½ cups of dry soybeans. Soak in 2 cups of water in an ice tray for 2 hours. Put in the freezing compartment of the refrigerator and freeze solid, preferably overnight. Remove from the refrigerator. Drop into 1 cup of hot soup stock or vegetable cooking water. Cover the container and simmer 2½ hours. Do not let beans boil. Add more stock if needed. Such beans can be put through a meat grinder, seasoned with pepper, garlic, salt, eaten as patties with tomato puree or used as a meat substitute in making meatloaf.

"Baked" beans can be easily prepared without the use of the oven, which is merely a convenient way of heating if you happen to have an old-fashioned range and the fire is going anyway. Actually the beans are not baked but slowly simmered. The lima bean makes a good basis for this dish, though kidney, navy or other beans can be substituted.

Wash and soak overnight in 1 quart of cold water 2 cups of lima beans. Simmer in same water ¾ hour. Fry ¼ pound bacon, cut in small pieces, remove when cooked and add 3 tablespoons chopped onion, cook gently till soft. Combine ½ cup molasses, 2 teaspoons chili powder, 1½ tablespoons brown sugar, 2 teaspoons salt, 1 teaspoon dry mustard, 1½ cup tomato sauce. Combine onion, beans, bacon and other ingredients, simmer 30 minutes.

A green vegetable can be prepared to balance this meal either from cabbage (fried rather than boiled) or from green peas, spinach or chard, depending on the season. Green vegetables should be cooked in the minimum of water or simply sauteed in oil. The summer squashes (zucchini, yellow crookneck, etc.) are especially good cooked in this way. A green salad, which even in midwinter can be prepared from bean sprouts or mustard and cress, will provide a sufficient level of vitamin C.

The Whole-earther who insists on devouring the bodies of his fellow mammals, be they deer, cows, pigs, sheep or rabbits, can get some valuable tips from Adelle Davis, *Let's Cook It Right*. Babies, she informs us, given free choice among meat dishes, go at once to the animal's brain and devour it raw with gurgles of enjoyment. A number of children given brain daily over a 5-year period starting in infancy had the highest intelligence ever recorded in the school they attended. The number, the author hastens to add, was too small to be statistically significant. She recommends the glandular meats (liver, spleen, kidneys, sweetbreads), which should be lightly cooked in oil to preserve the vitamins. No one can deny that liver, that great chemical factory of the body, is full of good things. Unfortunately, in this polluted world, it may also be full of bad things, for it is the liver that has the job of neutralizing toxic substances in the body. So even creatures as remote from civilization as the seals of Alaska may have dangerously high levels of mercury in their livers because they are the last link of a long food chain that progressively concentrates mercury.

FOOD STORING AND PROCESSING 111

The secret of successful meat cooking is temperature control. Adelle Davis gives some quite surprising facts about slow roasting. Meats which had been cooked for 26 to 32 hours were preferred in 100% of the taste tests to roasts which had been cooked for 3 hours or less. An oven temperature of 165° F is ideal for this type of roasting. Even a tough cut like the top chuck cut from neck muscles or heel of the round becomes tender enough to cut with a fork after 24 hours at 150–160° F. It is this slow, prolonged cooking that accounts for the tenderness of meats wrapped in grass or leaves and cooked in a fire pit. The temperature of the meat, however, must go above 140° F because this is the lowest temperature at which bacteria are killed. The trichinae in pork, which cause trichinosis, are killed in a few minutes at 131° F (55° C) so these are safely dispatched at 140° F. But be sure the meat really attains this temperature inside. To check this it is really worth investing in a meat thermometer, which can be stuck into the meat. It takes a surprisingly long time for a large lump of meat to attain oven temperatures. It will heat much more quickly if surrounded by steam, as in a pressure cooker or boiler.

The old-fashioned method of barbecuing meat over an open fire results in well-cooked meat. A revolving spit causes the meat to be evenly heated and as one part heats another part cools, resulting in a low average temperature within the meat mass. The meat surface becomes browned but the meat shrinks less than if it were roasted in a closed oven.

Cooking meat with moist rather than dry heat is more economical in fuel and can give satisfactory results if excessive temperatures are avoided. There are two moist-heat methods, braising and stewing. Braised meats retain their flavor because they are exposed only to steam, not to boiling water. Stewing is really a cook's last resort and should be used only with meat so tough it can be cooked in no other way. Slow braising, like slow roasting, can greatly improve the flavor of meat. The tenderness of commercially boiled ham results from prolonged cooking at 160–165° F. It may take 8–12 hours to prepare a pot roast by this method but the low heat results in a more tender product and is also economical in fuel. This slow-steaming method is especially valuable when solar cooking is used.

Meat can, of course, be considerably tenderized by the use of such enzymes as papain, which forms the basis of various commercial tenderizers. The owner of a papaya tree need only wrap his meat in one of the leaves, punch several holes with a skewer through the leaves into the meat and leave the enzyme to work overnight. Marinating meat in vinegar or lemon juice also helps tenderize it. Wild rabbit meat and that of other game soaked in brine for 24 hours will taste less gamy, though all game should be hung for about 3 days before using.

Meat and beans cooked together with plenty of chili (chile con carne) provide a powerful body fuel rich in protein, food fit for loggers or mothers-to-be (but not high enough in calcium for the latter). You can make it as follows:

Soak 2½ cups of kidney beans overnight. Pour off the water, rinse. Boil beans in fresh water until tender. Brown in hot fat 1 pound of ground meat and 1½ cups minced onions. Mix beans and meat, add 1½ cups tomato juice, blend in 1½ to 2 teaspoons chili powder, 1 teaspoon flour, 3 teaspoons water, 1 teaspoon salt. Let the whole brew simmer for 45 minutes in a Dutch oven. This can be the basis of various meat stews. If you don't like it hot, leave out the chili. If you like turnips, carrots, potatoes or rutabagas, cook them separately and throw them in. You can even leave out the meat and still get a nourishing stew by using soy curd or soy grits in place of meat.

Being myself a fisherman and relying on fish for my animal protein I pay special attention to fish cookery. The following points should be remembered about fish: They are wet. The flesh of fresh fish contains 65 to 80% moisture. If you want to get the same amount of protein from fish as from meat you must eat more of it. Fish is not high in fat. Fatty fish such as shad, mackerel and herring average 7 to 10% fat, salmon has 14% but other fish may have 1% or less. By contrast beef averages 33% fat and ham 50%. Ocean fish are rich sources of phosphorus, calcium, iodine and various B vitamins if these are not soaked out of the fish by careless washing. Fish liver contains much vitamin A and D, though the liver is rarely eaten. Fish roe, being largely nucleic acid, is rich in phosphorus. Shad roe is especially tasty but the roe of the cabezon is poisonous.

Fish cooks when it is little more than warm. An internal temperature of 140–145° F indicates that it is cooked. To boil fish is nothing less than criminal treatment of good food. The best method: deep fat at 360° F. If you cut your fish in boneless fragments you can coat it in batter made as follows: Sift together ⅓ cup of whole wheat pastry flour, 3 tablespoons powdered milk, 1 tablespoon mustard (optional), 1 teaspoon salt. Add and stir well ¼ cup fresh milk and 1 egg. Dip fish in the batter one piece at a time, place in a wire basket and fry in deep fat. The cooking is fast, 1–2 minutes depending on the size of the fish lumps. The result is fish-pops, a delectable dish. You can eat them hot or cold, carry them in a knapsack. They will keep several days. Even without the batter deep fat is the best way to cook fish. Carp loses its muddy flavor when cooked this way. Pan frying is next best if the fish is not too large. Larger pieces are best coated in flour, cornmeal or breadcrumbs and sauteed. This is a very effective method of cooking mussels. Dipped in egg and breadcrumbs these readily available bivalves are delicious.

A good way to use leftover fish is in *fritatas*. The delectable patties are made by boning the cooked fish, breaking it up and mixing it with breadcrumbs, ½ grated onion, 1 egg, garlic, soy sauce, 3 tablespoons parmesan cheese, and any leftover vegetable like zucchini, cauliflower, winter squash. Fry until brown. These *fritatas* are a meal in themselves. If you have no parmesan cheese, just leave it out, the fish provides the protein.

Cooking heat. In cooking, the question is: where do you get your heat from and how do you apply it? If you refuse to use coal, natural gas, oil, electricity or atomic power, then you have three sources left: the sun, waste vegetation (wood, bark, grass, pine cones, etc.) or gas derived from waste material.

He who would cook with sun power can manage very well if he lives in a sunny spot. Sun power is beautiful stuff, the light and life of us all. It comes for free in great daily surges of photons, a lot of it down in the infrared region, which good cooks feel broils best. But the photons are rather diluted (which is just as well or we'd be broiled alive), so the essence of solar cooking lies in bringing photons together and focusing them on the cook pot. Actually the direct focusing is not the most intelligent way to use solar energy. A better way is to use a heat storage device, which can be sand or soil in a black metal box. By sinking the reflector below ground level the cooker can be very neatly arranged to cook food quietly and steadily without any danger of burning the food.

Solar cookers can take three forms depending on whether they use parabolic reflectors, plane reflectors or lenses. A cooker using a parabolic reflector can be made from an army surplus search light. In case you cannot get one of these the parabolic mirror can be made, in segments, from aluminum sheet. A parabolic mirror with a diameter of 3½ feet and a focal length of 18 inches will focus enough heat on the base of a Dutch oven to cook almost anything, provided the mirror is adjusted from time to time to compensate for the earth's movement and the altered angle of the sun. It is possible also to build a parabolic mirror in the form of a trough about 4½ × 9 feet. The formula of curvature of such a mirror is $y^2 = 120x$.

Simpler to make is a plane reflector cooker of the type developed by Dr. Maria Telkes. It involves eight flat sheets of polished copper or aluminum mounted on a hinged frame at such an angle that the sun's rays are directed onto an oven in the center of the array. The oven should be black to absorb the heat. With this device oven temperatures of 400° F can be attained (fig. 14).

A still simpler arrangement is to heat a fire pot by means of a Fresnel lens. These flat plastic lenses can be obtained from Edmund Scientific Co., 150 Edscorp Building, Barrington, N.J. 08007. The 19½ × 24¾-inch lens can be arranged to throw solar heat on an insulated fire pit of sand or onto an insulated oven. If the lens is adjusted to focus more sharply it can be used as a solar furnace.

FOOD STORING AND PROCESSING 115

All these devices using sunlight will indeed cook a meal if the sun is shining. For days when it isn't, alternate forms of heat are needed. The fire pit is simple but it has obvious drawbacks. It would be hard to bake bread in a fire pit. But if, using stone or adobe brick, you raise your fire pit off the ground and then close it in with an arch of brickwork you have an earthen oven in which waste wood can be burned. Light a good, hot fire in the pit, rake

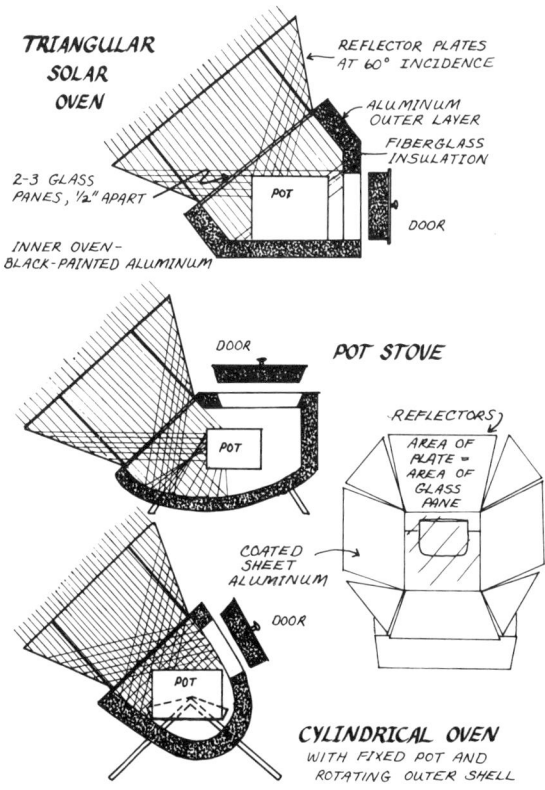

SOLAR COOKERS (Maria Telkes Design)

FIGURE 14

out the coals, put your loaves in and place a sheet of metal over the opening. Pile the hot coals against the metal. The bread will bake slowly and evenly. If you have enough space and are handy with adobe you can add to the oven a range, with a water heater to utilize waste heat. A good base for such a stove is a heater made from a 50-gallon drum, the ironware for which can be obtained from army surplus stores. The drum, if properly insulated, can itself be used as an oven, provided all the coals are raked forward and the loaves are put in the back. This arrangement is fine for space heating as well as cooking. In warm climates, of course, the extra heat is a drawback and the whole assemblage should be placed outside the building.

All this can be replaced by the old-style kitchen range and these cast-iron monsters are still being built (try Portland Stone Foundry Co., Box 59, Sterling Junction, Mass. 01565). But they cost a lot new (over $600 retail) and it is more instructive to build one's own stove out of materials readily available (such as mud).

The finest single utensil for cooking is the cast-iron or aluminum Dutch oven. With one of these or a pressure cooker one can cook an adequate meal with nothing more elaborate than a fire pit. A fire in a pit heats the surrounding rocks and the rocks hold their heat a long time. A Dutch oven, filled with suitable food and firmly closed, can be placed amid the hot rocks or in the hot sand of the fire pit, and the food left to simmer. This long, slow cooking is very satisfactory for beans and for meat.

You can get a Dutch oven from Lodge Manufacturing Co., South Pittsburg, Tenn. 37380. This, along with a pressure cooker and a deep skillet, is all the Whole-earther will need. He doesn't really need the skillet because the lid of the Dutch oven can be used for frying.

A modification of the fire pit and hot stone technique is the fireless cooker. This is simply an insulated box into which is placed a specially molded rock or concrete slab that has been heated in the fire. The insulated box can be cylindrical or square. A metal garbage can is quite satisfactory. Straw, hay, shredded newspaper, or fiber glass insulation as used in buildings is put into the box as a liner. The inner compartment (well) can be made from sheet metal. The flat stone that carries the heat should be cast in cement with a wire handle for lifting. If you live in a tipi your fireless cooker can be sunk in the ground next to the fire pit.

To use the cooker, first heat the heating stone by setting it in the fire. Transfer the heating stone to the cooker, then put the covered cooking pot on the stone and place the insulated lid on the outer jacket. Cereals like rice, cracked wheat and oatmeal should be brought to the boil for 5 minutes, then placed in the cooker. Beans should be soaked overnight, boiled for 5 minutes, placed in the cooker for 4 hours. There is very little loss of water in the cooker so a minimum of water should be used.

II SHELTER

Food is man's first need, shelter his second. He is a naked ape, left by nature poorly protected. He does not have fur, feathers, or the layer of blubber that serves for such mammals as seals. He is the only primate who lives in cold climates. The monkeys and apes all congregate in the tropics or subtropics. But man can survive even in the Arctic because he long ago learned to make what nature failed to provide. Having no fur himself he learned to steal the fur of others, using skins of the beasts he killed to cover his naked body. Clothes were a form of shelter and they had the advantage of being transportable. Man moved around in a clinging tent of skins. He carried his own personal climate, warm and cozy.

CLOTHES

1. Skins

One does not have to go all the way back to the Neolithic to find out how to clothe oneself in skins. The Eskimos are doing it now and have been for centuries and what they don't know on the subject is not worth knowing. An animal skin has two parts, the skin itself and the hair attached to the skin. The best protection from cold weather, especially wind, is obtained by wearing the skin outside and the fur inside. This effectively insulates the wearer with a layer of air which is warmed to the temperature of the body.

To prepare a skin with hair attached (such as sheepskin), first remove the skin from the animal carefully to avoid damage. Next lay out the skin on a flat surface and scrape off all adhering flesh and fat. This process is facilitated by soaking the skin overnight in salt water. When the skin is completely free of fat and flesh it can be worked either by pounding with a hammer to force the natural oils into the skin, or if you want to follow Eskimo usage, by chewing it until it softens.

A more advanced method of preparing whole skins involves pickling. After the skin has been soaked, remove not only adhering flesh and fat but also the soft areolar tissue which is part of the dermis, or true skin. This is done by drawing the skin over a razor-sharp knife supported in a suitable holder. Prepare a pickle by dissolving 1 quart of salt in 1 gallon of hot water. Let it cool. Slowly add, with stirring, 1 fluid ounce of commercial sulfuric acid (*handle with care!*). These are the proportions only. The total amount will depend on the size of the skin and of the vessel used. Do not use a metal vessel. The acid will attack it. An old wooden barrel with one end knocked out is suitable. Soak the skin, making sure every part makes contact with the pickle. A thick skin will be tanned in about 2 days, a heavy one will take a week. It will not hurt to let the pelt stay in the pickle for weeks. It does not injure hair or fur but sets it and discourages attacks by moths.

Faster pickling will be obtained by using 2 quarts water, 1 pound salt, 1 ounce of sulfuric acid. It will tan a light skin in about 24 hours.

After removing the skin, go over the flesh side with a scraper to remove surplus liquid and neutralize the acid by soaking the pelt for an hour or so in a solution of washing soda (about a handful to a pail of warm water). Rinse in clear water. Complete neutralization of the acid is important.

Soften the pelt by hanging it up until it is half dry, then work it back and forth, skin side down, over the edge of a plank and pull and stretch it in all directions until it is white, dry and supple all over. A final finish can be put on the dry skin by rubbing with sandpaper or pumice stone. Rub into the skin a mixture of equal parts of tallow and neat's-foot oil. Remove excess grease by rubbing with dry sawdust. Comb out the fur and the pelt is ready for making into a garment.

If you want skin without the attached fur, soak the skin overnight in a slurry of 30 pounds slaked lime in 10 gallons water. Scrape the hair off, along with the epidermis, which lies under the hair. When hair and flesh have been removed you can prepare rawhide by laying the dry hide on a pad of grass or an old blanket and pounding it with short, glancing blows from a hammer weighing 3 or 4 pounds.

A primitive but effective way of tanning rawhide involves making a slurry from the animal's brain. Boil the brain, press the material through a sieve, stir it back into the broth and soak the hide in the slurry overnight. Wring it out, stretch and knead until completely dry. To give the skin the golden tan associated with buckskin, sew the skin into a cylinder, attach a short skirt of cloth to the bottom and suspend over a bucket containing smoldering wood chips (preferably willow). Keep watch lest the chips burst into flame.

Tanning with oak bark involves grinding the bark, making a slurry of 5 pounds of bark to a gallon of water. The hide has to be left in the weak liquor for as long as a month, then moved into a stronger oak liquor and finally into the strongest of all. Rate of tanning depends on the warmth of the weather. Dubbin or tallow is rubbed into the hide to soften it after drying. Black leather can be made by adding ferrous sulfate (copperas) to some of the tanning liquor.

Skin clothing. Untanned rabbit skins were used by the Paiutes to make the rabbit skin blankets which were a vital part of their winter apparel. The skins were taken from rabbits killed in November. Rabbits were skinned carefully to keep the whole pelt intact. The incision was made along the inside of the hind legs and the whole skin pulled down over the body like a glove. Flesh and fat were scraped off. Beginning at the eye hole the skin was cut into a thin spiral strip 10 to 15 feet long. Each strip was formed into a loop by tying the tip of each fur ribbon into its eye hole. The loops were linked together to form a chain 40 or more feet long, which was then twisted into a rope. The skin of the rabbit folded in, leaving the soft fur on the outside. The rabbit skin rope was hung up to dry and the brittle ears were snapped off. Then the rope was woven into a blanket with strands of hemp. A man's blanket required 100 rabbit skins, a child's 40.

Clothes can be made either of leather or fur by cutting out the pieces according to the pattern and sewing them together. Fur should be cut with a knife on the leather side and sewn with an over-and-over stitch (Polish fur stitch). A small thread doubled may be better than a single thread and by passing the thread over a lump of beeswax it can be made to run more smoothly. It should be remembered when making fur garments that the Eskimos, who know most about this sort of clothing, always put the fur inside as this gives better insulation.

A buckskin shirt will require two large hides or three small ones. For a pattern it is best to use a comfortably fitting old shirt, take it apart at the seams, dampen the pieces and press them flat, then cut out the pattern. Sew inside out and turn. Be careful to leave plenty of room across shoulders and chest. For the fringe across the shoulders, cut a piece 2½ to 3 inches wide, cut the fringe to 2 inches deep and sew into the shoulder seam. A three-cornered needle can be used for this work and linen carpet thread well lubricated with beeswax. Do not make a shirt which opens down the front like an ordinary shirt but leave a shoulder seam open from the neck to the top of the shoulder and insert a zipper or leather lace to close. Properly smoked buckskin can be washed in tepid soapsuds and will dry soft.

Moccasins can best be made by combining a heavy leather sole with more flexible uppers, a technique used by the Sioux. The piece of heavy hide is cut to the shape of the bottom of the foot. This can best be sewn to the uppers by using a "waxed end," which is made by running several strands of linen cobbler's thread through a lump of cobbler's wax and rolling them on the knee into a single thread. The wax stiffens the ends. Holes are made with an awl and a double stitch is used, one thread inserted through the awl hole from above, the other from below. This makes a strong, tight bond which is more or less waterproof.

2. Cloth

One not willing to dress in skins will need some kind of cloth from which to make clothing. The making of cloth takes us back through the centuries to those men of the Neolithic who discovered spinning and weaving. Spinning consists in twisting vegetable or animal fibers into thread from which cloth can be woven on a loom. The true Whole-earther can grow three forms of fiber, all suitable for clothing. These are flax, wool and cotton.

Any Whole-earther with a little soil to spare can experiment with flax if he/she lives in a temperate climate. The flax plant (*Linum usitatissimum*) with its fragile blue flowers and slender stem is a charming plant and easy to grow even on infertile soil. The art of making linen is so ancient (it attained a high level in ancient Egypt and was spun so finely as to give almost transparent fabrics) that the process is an archetypal experience. Allow your flax plants to mature until the spherical seed heads are beginning to turn brown (fig. 15). *Pull* the plants up by the roots and bind them in bundles. Make a coarse *rippling comb* by driving large nails through a board about half an inch apart. Ripple the flax by pulling it through the comb. This removes the seed pods, which can be dried, stomped and winnowed. The seeds are rich in linseed oil. Next *ret* your flax. Lay the stems in a pool of water weighed down with a rock. Leave them for 10 days to 2 weeks. The outer straw and inner pith will partly rot. The fibers will not be affected unless you overret, in which case they too will begin to rot. Next *grass* your flax, laying it out on the grass in the sun to bleach. When one side has bleached, turn it over and bleach the other side.

The flax is then ready for *scutching,* a process which breaks the stems and makes possible the removal of the core and outer parts of the plant. To do this effectively, first make sure that the flax is absolutely dry and work in a warm room to keep it dry. Prepare a board with a slot in it and hold in the slot a handful of flax by the root ends. Beat the flax against the side of the board with a wooden blade. Twist the bundle during the process, then reverse it so that the root ends can also be beaten. In this way the *boon* is removed and the flax made ready for *hackling.*

Hackling is the final process in dressing the flax. Take a bundle

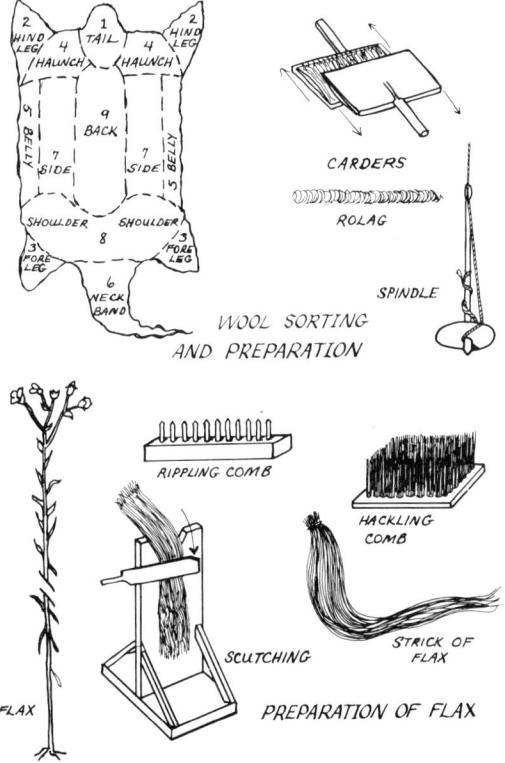

FIGURE 15

of fibers by the root end and pass them through a coarse-toothed metal comb. This removes the rest of the boon and the short fibers. Repeat this process through a comb with finer teeth. The short fibers which are hackled out are called *tow,* the long strands are called *line.* A bundle, or *strick,* of flax that has been properly hackled looks like a mass of well-combed blonde hair.

When flax is spun and woven, the product is linen, a perfect fabric for certain purposes, but one which lacks warmth. For warm clothes the true Whole-earther will turn to wool. To understand why wool is warm one must look at its structure. Open a fleece and remove some of the wool. You will notice that the wool grows on the sheep in definite locks. These locks are the *staple.* Length of the staple depends on the breed of sheep. Coarse, long-staple wool grows on Mountain breeds like the Scottish Blackface and is more suited for making rugs and carpets than clothes. Short, fine-staple wool grows on the Down breeds like the Southdown and especially the Merino.

Next look at the staple and observe the *crimp.* These fine waves in the wool fibers give wool its warmth. They ensure that even when wool is tightly woven it still holds entrapped air. Moreover, the crimp remains even when the wool is wet, which accounts for the fact that even a wet woolen garment is warm. Notice also that wool in the fleece is not all of the same quality. Indeed, wool from various parts of the fleece is so different that fleeces are always pulled apart and separated into wool types before being processed. This is the job of the wool sorter, practicing a difficult art learned only through long experience.

The true Whole-earther who wants to enjoy the full wool trip (and a very ancient archetypal trip it is) will first have to decide what sort of wool he wants and what breed of sheep will thrive in his part of the world. A hardy Mountain breed like the Scottish Blackface can survive in the Highlands and the Islands (that is, the Hebrides) where weaker breeds would starve. Their long, coarse wool is fine for carpets and can be used for Harris tweeds. The Corriedale from New Zealand is a cross between the Lincoln and the Merino and is becoming popular in California. The wool is dense, lustrous and very long in staple—up to 26 inches. The Kent or Romney Marsh is noted for its ability to stand cold driving rain and to thrive on marshy soil without developing foot rot. Its wool is close and fine with more definite crimp than most long wools. It has good felting qualities and is very useful to the hand spinner. The Southdown is one of the best known of the Down breeds. It yields excellent lean meat and fine, short wool very close and very crimped.

Having chosen his breed sheep the Whole-earther must next learn how to shear it. Lambs are first shorn at the age of about 15 months in late spring or early summer. The small operator can best do his shearing with a hand-operated clipper such as is used for grooming horses. Start by removing the belly wool and keep this separate. With the sheep sitting up on its rump, remove the wool on the left side with short, circular movements, then lay the sheep on its side and remove the right-side wool with long strokes from rear to front. With Merinos special care must be taken not to damage the neck folds.

The fleece holds together and can be rolled up and secured with the neck wool. To sort it, lay it out on a large table or on the floor and tear it apart as shown in fig. 15. Seven piles of wool will result: (1) Shoulder wool, fine, well crimped and strong. (2) Belly wool, finer than 1, well crimped but tender. (3) Back wool, coarser, harsher, less strong than 1. (4) Britch wool from haunches and tail, strong, coarse and hairy and frequently a bit stained. (5) Side wool, longer, straighter and slightly less fine than 1. (6) Foreleg wool, short and coarse. (7) Under-neck wool, fine but rubbed and felted.

Careful sorting is essential. Different parts of the fleece have different properties. The best wool is class 1, with 3 and 5 next. Belly wool tends to be weak. The only way to learn wool sorting is to do it, preferably under the guidance of an expert. The different classes of wool should be stored unwashed in mothproof bags until ready for spinning.

Wool to be spun must first be teased. Take a handful of fleece and pull the locks apart so that straw, seeds, etc., fall out. Teasing converts the compact lock into a fluffy, airy cloud of wool with no dense patches. This treatment alone is sufficient to prepare wool for spinning on a spindle. Before spinning, add a few drops of olive oil to the teased wool.

It is worth the Whole-earther's while to begin spinning with a spindle rather than with a wheel. Using a spindle is an archetypal experience which links the spinner to a host of others, all of whom have used this method of generating thread. A spindle, moreover, can be easily made by the spinner or bought for a dollar or two, but to make a spinning wheel is much more difficult.

The spindle is simply a straight wooden axis with a circular whorl at the bottom to give it momentum. The whorl must be easy to remove. The tapered axis must be notched at the top like a crochet hook. A length of coarse woolen yarn is first threaded on the spindle (fig. 15). To this yarn the teased wool is added by twisting the spindle in a clockwise direction. Draw out with the left hand enough fibers to make the yarn and release the twist from the right hand so that it will travel up and make strong yarn. Keep the spindle revolving in the right direction. It will tend to go into reverse. If it does so, it will *unspin* and the yarn will fall apart.

Winding on involves slipping the yarn from the notch at the top of the spindle and removing it from under the whorl, then winding it crosswise up and down the spindle. Leave enough unwound to hitch it to the spindle as before.

Yarn spun in the clockwise direction is known as Z-twisted. A good spinner will also learn to spin in an anti-clockwise direction, changing the role of the right and left hands. Yarn spun in the anti-clockwise direction is known as S-twisted. Finally, the spindle must be emptied, which is done by slipping up the whorl and transferring the cone of wool to a stand made by inserting four knitting needles into the corners of a square of plywood. From this the yarn can be wound into a wrap reel on which it can be tied in skeins. Skeins must be carefully tied so that they do not come apart in the dye bath.

Wool to be spun on a wheel (fig. 16) must be *carded* as well as teased. The teased wool is laid on one of a pair of wooden bats called carders, brushed out, transferred to the other card and so on back and forth until it can be rolled off the card in the form of a loose cylinder called a *rolag*. If you wish to work with long wool fleeces you can prepare the wool by *combing* rather than carding. Take a lock of wool 4 inches in staple or longer, put a drop of olive oil in the palm of one hand, roll the lock between two hands to distribute the oil. Clamp a metal comb to a table and, holding the prepared lock by the cut ends, draw it through the comb several times, then turn it over and comb the underside. Next take the lock by the tips and repeat the process.

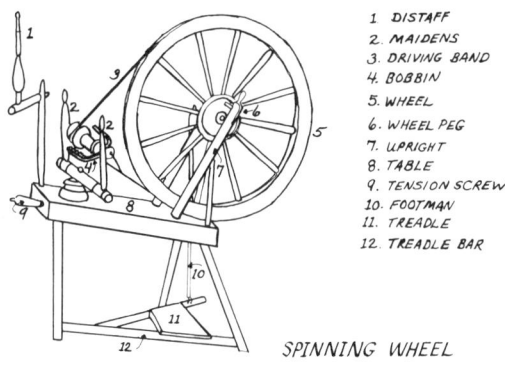

SPINNING WHEEL

1. DISTAFF
2. MAIDENS
3. DRIVING BAND
4. BOBBIN
5. WHEEL
6. WHEEL PEG
7. UPRIGHT
8. TABLE
9. TENSION SCREW
10. FOOTMAN
11. TREADLE
12. TREADLE BAR

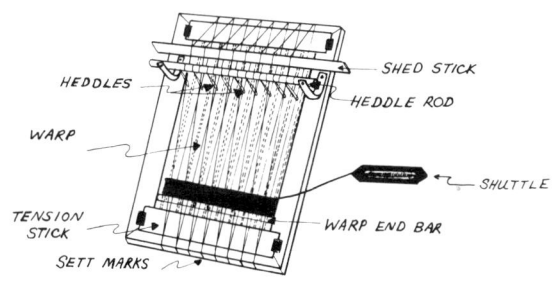

SIMPLE FRAME LOOM

FIGURE 16

Short fibers called *nails* will be collected in the teeth of the comb. The long fibers called *tops* will remain, straightened out and parallel, in the hand.

Such long-staple, silky wool is suitable for a form of spinning called *worsted*, as opposed to *woolen* spinning which gives a closer, more felted yarn.

Both linen and wool need further treatment after spinning. Wool may be woven "in the grease" and scoured later or scoured and dyed before it is woven. To *steep* wool the skeins are placed in hot (50–55° C) soft water or rain water and allowed to steep until the water is cold. Next the wool is soap-scoured by being placed in warm, soapy water (45–50° C). The yarn should float freely. Turn skeins over once or twice but handle them as little as possible. The quickest way to make wool felt up is by squeezing it in hot, soapy water. When the skeins are sufficiently scoured, rinse them until the water is clear and hang them up to dry. Woolen-spun yarn can be dried without tension. Worsted-spun yarn should be stretched until dry.

Linen yarn will vary in color, depending on how the fiber was retted. To soften yarn, boil it in a solution of soap for an hour or so, cool it and rub the yarn vigorously; rinse and repeat the process until the yarn is soft. To bleach linen yarn, lay the skeins out on the grass in the sun, keeping them damp and turning them occasionally. For more rapid bleaching, use a dilute solution of Clorox in an earthenware, glass or wooden container. When the desired shade is obtained, wash out all residual bleach and hang skeins up to dry.

Dyeing. Wool can be dyed in the skein or woven first and then dyed in the piece. If you use chemical dyes there is not much art involved. The instructions come with the dye and the results, if you follow the instructions, are generally predictable. If you use vegetable dyes you confront a different problem. You can either gather your own, in which case you are confined to the local flora, or buy the plant material. The number of natural dyes is large and the science or art of dyeing is complicated by the fact that different mordants give different colors with the same dye. For detailed information, read *Vegetable Dyeing* by Alma Lesch, or *Dye Plants and Dyeing,* Brooklyn Botanic Garden, both available from *Last Whole Earth Catalog.* A complete collection of natural dyes can be obtained from Dominion Herb Distributors, 61 St. Catherine Street West, Montreal 18, Que., Canada. For other types of dyes, try Fezandic and Spearie Inc., 103 Lafayette St., New York, N.Y. 10013.

Weaving. Like spinning, this is an ancient craft and an archetypal experience. The true Whole-earther who wishes to enjoy this experience should first familiarize himself with its language. Weaving involves the use of a set of parallel threads (the *warp*) strung tightly between two supports. Between these threads a second set of threads is inserted at right angles (the *weft*). In order for the weft to be inserted, the warp threads must be separated to form a *shed.* This is done by pulling one set of threads up and another set down, an operation performed by the *heddles.* The *reed,* attached to the *beater,* is made like a comb of polished metal. All the warps pass through the reed. The weft is carried on a *shuttle* that is thrown from one side of the cloth to the other and travels on a *shuttle race.* Each time the shuttle is thrown, the beater is pulled forward and jams the weft firmly in place. The position of the heddles is then changed by means of *treadles* attached to the heddles and worked with the feet. The warps are held on a roller called the *warp beam* and are carried over a heavy, smooth *back beam* to an equally heavy, smooth *breast beam* and the finished cloth is wound onto the *cloth beam.*

As a full-scale loom is an expensive instrument, the beginner is well advised to familiarize himself with the art of using an inkle loom, which can be obtained for as little as $10 (The Wind Bell Inc., 5714 Kennett Pike, Centerville, Del., or Good Karma Looms, Route 1, Park View Terrace, Chadron, Neb. 69337). Or build your own (fig. 16). These small looms do not have a reed and the heddles are hand-operated so work is slow, but those who wish to learn by doing (and what other way is there to learn?) can weave very fancy belts on an inkle loom and then, if the urge to weave becomes imperative, buy a full-scale loom (Gilmore Looms are recommended, 1032 N. Broadway, Stockton, Calif. 95205) or build one (for plans, see *To Build or Buy a Loom* by Harriet Tidball, Craft and Hobby, Box 626, Pacific Grove, Calif. 93950).

A primitive but effective instrument is the loom used by the Navahos for weaving blankets, rugs and ceremonial belts. This can be a permanent installation, for the heavy uprights are often buried in the ground and used as supports for a shelter as well as a loom. A single heddle stick can be used to move the warps or they can be opened with a wooden batten stick several inches wide and with one blade-like edge that can be inserted between the warps in any desired pattern and the weft introduced by hand. Instead of a reed a heavy beater with metal comb attached can be used to beat down the weft and make a firm fabric.

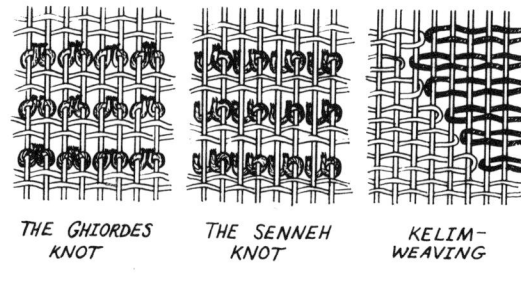

THE GHIORDES KNOT   THE SENNEH KNOT   KELIM-WEAVING

CARPET KNOTS

FIGURE 17

On the same type of loom as used by the Navahos a true Whole-earther with an interest in Oriental rugs can weave genuine knotted rugs of the Persian or Turkish variety. The knots are made across two warps (fig. 17) and two wefts are woven over each row of knots, the whole being beaten down with a heavy beater.

HOUSING

Housing is an extension of clothing, a device for providing a controlled environment. Foxes have holes and birds of the air have nests, bees have hives and termites have termitaries. Man has a house. On that house he is apt, at least in these United States, to spend more money than on any other single purchase. It becomes far more than a device for keeping out rain, snow, wind and various pests. It fastens on the man, becomes a symbol of his ego. It can turn into a veritable Moloch devouring vast sums of money. Many a man in this affluent society has sacrificed health, peace of mind and solvency to acquire some architect-designed monstrosity that was neither practical nor comfortable, but merely served to inflate its owner's sense of importance.

Ken Kern, who is a true Whole-earther, and the author of *The Owner-Built Home* (available from *Last Whole Earth Catalog,* or Ken Kern Drafting, Sierra Route, Oakhurst, Calif. 93644), makes the following points:

(1) Stupid, antiquated building laws may result in a waste of as much as $1,000 per house. Try to locate outside of urban control.

(2) A building loan is a type of legalized robbery. A long-term mortgage may double the cost of a house. Build what you can afford and pay as you go.

(3) The general contractor is an expensive and non-essential luxury. Once people realize how easy it is to implement a set of house plans they will dispense with the contractor.

(4) "Most contemporary architects design houses for themselves, not their clients. They work at satisfying some esthetic whim." A truer word was never spoken! Design your own house to fit your own needs.

(5) Use good hand tools. Overdependence on power tools involves the home builder in expense and prevents him from developing manual skills.

A dwelling can be temporary or permanent. The tipi, the yurt, the tent are all movable structures and the people who use them are movable people, nomads. As for the gypsy, he, in England at least, traditionally lives in a horse-drawn caravan, or *vardos,* gaily painted with traditional designs. The contemporary nomad may follow the gypsy's example, substituting a bus or truck-drawn trailer for the *vardos*. Having lived in a trailer through a winter in New York State I can testify that this form of shelter leaves much to be desired. Unless it is extremely well insulated, moisture condenses on the walls in cold weather and everything tends to get damp and grow molds. It does have the advantage of being totally self-contained. So does an old bus. So does a cab-over camper which any half-way handy man can build from a kit (for prices, write Luger Industries Inc., 1300 E. Cliff Road, Burnsville, Minn. 55378). But the motorized dwelling, unless run on methane from wastes, is not a form of shelter that the true Whole-earther would select. Its exhaust fouls the air, its cramped quarters confine the spirit. And the wandering life makes impossible any real self-sufficiency. One merely turns into a tramp, dependent for one's food on other men's labors.

There was nothing trampish about the genuine nomad, who now survives only in a few remote corners of the world. He was a true Whole-earther who took his sustenance from the biosphere via his flocks or via the animals he hunted. His austere pattern of life fostered toughness and independence. He followed his flocks from the valleys in winter into the mountains in summer, as the nomadic Mongols still do. Or he followed and preyed upon the buffalo herds, as did the nomadic Sioux. The challenging life, the sparseness of his possessions, the constant exposure to nature in all its moods, gave rise to a hard, practical, often ruthless human being who tended to despise the comforts of the settled life. It was these nomads who, under the leadership of Genghis Khan, became the scourge of civilization.

So the truly nomadic life, which involves following one's herds over wide expanses of unfenced pasture, is certainly one which no true Whole-earther can despise. It is, however, almost impossible to live such a life in America unless you follow your sheep or herds of cattle through those national forests open to grazing. For such a one the nomad's dwelling in its ancient or modern form may be just what is needed. If he wishes to avoid depending on a motorized conveyance he must reduce his needs to those which can be carried on the back of a horse. This includes his shelter.

1. Tents

The most compact shelter for a true nomad is a tent. He can choose between the Warmlite designed by an aerodynamicist, 10 feet long, 5 feet wide, weight 36 ounces (Stephenson's, 23206 Hatteras St., Woodland Hills, Calif. 91364); the Bishop Ultimate 4-man (Bishop's Ultimate Outdoor Equipment, 6804 Milwood Rd., Bethesda, Md. 20034) or a 6½-foot green pup tent folding into a bag 5 inches × 2 feet and weighing 13 pounds (Whole Earth Truck Store).

Or the nomad can make his own tent, using the design of the yurt but omitting the smoke hole. The yurt is circular. The tent can be hexagonal or octagonal, depending on the number of supports you wish to use. It can operate on the principle of the umbrella with spokes radiating from a central pole, or the supports can come together from the six corners and tie in at the center. The supports for such a tent can be made of bamboo or light metal tubing with rubber tubing at the bends and suitable spacing pieces to give the frame rigidity. The segments of canvas can be cut and sewn together and the floor sewn in to give a single waterproof unit.

An even simpler shelter is the tube tent, which is nothing more than a tube of plastic or canvas held up by a cord strung between two supports. Though it is open at both ends and therefore well ventilated it does keep off the rain and is the lightest of all forms of shelter.

2. Tipis

The yurt and the tipi, both of them used by nomads for centuries, suffer from one drawback. They are quite bulky and hard to carry around. The poles for a tipi are at least 20 feet long. The wooden members upon which a yurt is constructed, though much less clumsy, are nonetheless bulky and hard to carry. So yurts and tipis are not really very practical dwellings for the modern nomad. They actually seem more suited to the settled Whole-earther whose romantic nature calls for this rather exotic type of shelter.

Reginald and Gladys Laubin, who built tipis, lived in tipis and knew more about tipis than many of the Indians, have written the definitive work on this subject (*The Indian Tipi,* Ballantine Books Inc., 101 Fifth Ave., New York, N.Y. 10003). The following much condensed account is taken partly from their book and from an excellent article "The Plains Indian Tipi," in *The Mother Earth News,* vol. I, no. 1, p. 29.

(1) Poles. For an 18½-foot-diameter tipi, make 17 poles about 25 feet long, 3 to 4 inches thick at the butt, 2 inches where they cross and tie. Red or white cedar, lodge pole pine or yellow pine make good poles. Cut trees early in spring, trim off all branches, peel off bark with a sharp drawknife. Season poles by laying them flat across pieces of scrap lumber spaced 2 feet apart. Let poles air- and sun-cure, turning regularly. Finish with a few coats of linseed oil. If you cannot find suitable trees, buy first-quality 2 × 4's, rip them on a taper and round edges with a drawknife.

(2) Cover. The Indian tipi was covered with buffalo hide. The modern Whole-earther will certainly not find buffaloes and will have to do with 8- or 10-ounce duck canvas. The 72-inch material saves work but is hard to find. The 36-inch material can be obtained readily. The 8-ounce is easier to work with, makes a lighter tipi and is cheaper. Cut six strips as shown in fig. 18 and sew them together firmly and forcefully using double seams. (This is best done by hand with carpet thread coated with beeswax. It's rough on a sewing machine.) When the strips are sewn, lay out the cover, locate the center of the longest strip. Cut off two 20-inch-wide segments from the center point. Sew these two segments together and add them to the cover at the bottom. Draw a semicircle, with chalk and string, of 19-foot 3-inch radius. Mark the center and cut as shown in fig. 18 to make the smoke flaps and door opening. It is nice though not essential to trim the door ovals with soft leather. Cut and stitch the lacing holes between the smoke flaps and door opening. They can be spaced 4 to 7 inches apart.

To shape the tie flap so that it fits snugly around the poles, cut a 39-inch opening as shown in fig. 18 and sew in a triangular gore 39 × 39 × 7 inches. Sew an 8 × 24-inch extension onto each smoke flap.

HOUSING 135

Sew pole pockets to the top corner of each smoke flap. Sew two tapes, 3 feet long, to the tie flap and an 18-inch tie flap to the base of each smoke flap. Make the tapes by folding together a 3-inch-wide strip of canvas into a triple-thick, inch-wide band that is double-stitched down both sides. The tape on the base of the left smoke flap is sewn to the top side of the hem, and that on the right smoke flap is sewn on the underside of the hem. Buttonhole-stitch a hole in the lower corner of each smoke flap and attach 6 feet of 3/16-inch cord.

To complete the tipi, make a liner out of heavy bleached

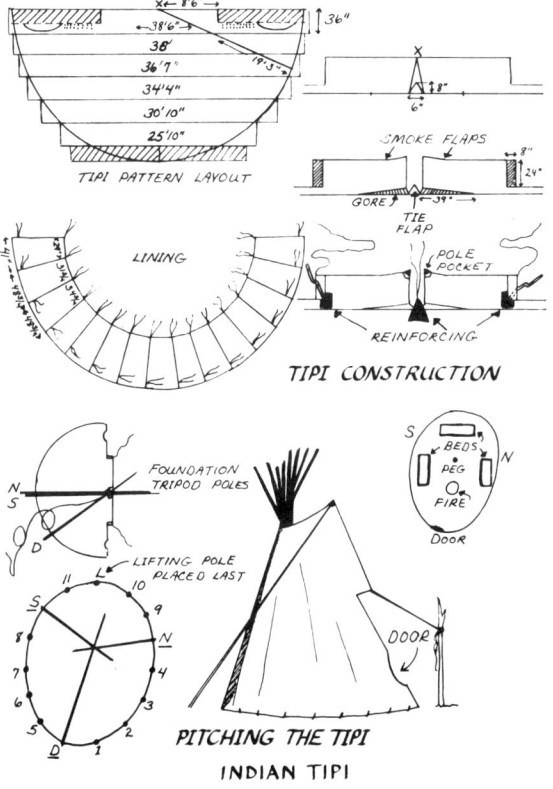

FIGURE 18

muslin, 8-ounce canvas or some of the coated fabrics available for tent making. The liner hangs around the inside of the tipi and creates a tent within a tent. It is important that the liner be waterproof. It can be made from 15 identical tapered panels each 6 feet long, 34¾ inches wide at the top and 48¾ inches across the bottom. Sew the panels together, reversing every other one. The long strip of fabric so produced should go completely around the inside of the tipi and lap generously. A double cord, 3/16 inch, sewed top and bottom to each vertical seam will complete the liner. The ties should be 24 to 30 inches long and the lower ones should be 6 to 8 inches from the lower edge to enable the liner to be turned in all the way around. The top of the liner is tied to a separate rope running around the tipi poles. This prevents drips from running down the liner. The bottom of the liner is tied to the butts of the poles.

To pitch a tipi, select a site that is smooth and level. Pick out four of the heaviest poles. Three will be used for the foundation tripod and the fourth will be the lifting pole. Spread out the tipi cover and use it to measure the tripod poles (fig. 18). Allow a few inches for that part of the pole that will be sunk below ground level. Mark the three supports where they cross. Take 45 feet of ½-inch manila rope, tie the three poles with a clove hitch, wrap the rope around the poles three or four times and finish with two half hitches. Because the prevailing winds on the Plains were from the west or northwest the tipi was pitched with the entrance facing east. Poles labeled N (north), S (south) and D (door) are therefore spread out as shown in fig. 18. The tipi is a tilted cone and the top of the smoke hole slopes away from the door. After the tripod has been raised, put the other 11 poles in position following the order shown. Leave a gap at L for the lifting pole. Wrap the rope four times around the standing poles, bring it over the point where the poles cross and anchor it to a peg driven into the ground in the position shown.

Lay the lifting pole down the center of the cover and mark it at the tip of the tie flap. Remove the lifting pole and fold the outside edges of the cover to the center. When the cover is neatly folded, tie it to the pole at the position marked, hoist the pole and canvas and drop the butt of the pole into the correct position, rotating it so that the cover is on the outside. Spread out the cover, tie the smoke hole base tapes together and insert the top lacing pins. Insert poles in the smoke flap pockets. Peg down the edge of the cover, using peg loops made of 3/16-inch cord tied around a pebble or marble in the cover. Pegs should be 27 inches apart. Prepare the fire pit, which can be ventilated from below by means of flue pipe buried in a trench leading outside the tipi. Hang the lining inside and turn in the bottom 6 inches. Cover the floor with buffalo skins if you want the authentic effect, or use linoleum, tarpaulin, plywood, oilcloth.

Ditch around the tipi if you wish to lead water away. Make a couple of willow rod backrests. Hang your shield and other medicine articles from the pole behind the altar. Hang a door (canvas on a willow frame) over the opening and move in. If the weather gets too hot, lift the skirts of the tipi. If too cold, build a windbreak of straw around it. Paint the cover with the design you fancy.

As for the romance of tipi living, let the Laubins speak: "It is a joy to be alive on days like this, and when we come back to the tipi, after a long ride or hike into the mountains, the little fire is more cozy and cheerful than ever. The moon rides high in the late fall nights and when it is full shines right down through the smoke hole. Its pale white light on the tipi furnishings, added to the rosy glow of the dying fire, is beautiful beyond description."

Of course, when it's pouring rain and water is dripping from the lodge poles one's emotions may be somewhat different.

As for those that have not the time or the skill to make their own tipis they can buy them ready-made from Nomadics, Star Route, Box 41, Cloverdale, Ore. 97112. These are excellent tipis and very reasonably priced.

The yurt is simply a flattened tipi and approaches a dome or igloo in outline. The lattice-work of sticks covered with sheets of felt can be assembled relatively easily by Mongolian nomads who have been doing this all their lives but it is certainly more complex to build than a tipi and offers few advantages. Like the tipi it has a hole in the roof which lets smoke out and also most of the warm air. It *is* circular, which seems to satisfy an essential need in some people, but if you're going to build a yurt you might as well modify it; make it more or less into a dome, get rid of the smoke hole and install a fireplace with a chimney that can at least be closed when not in use. Bill Coperthwaite is an authority on yurts and offers a yurt construction plan ($3.00, William S. Coperthwaite, Bucks Harbor, Me. 04718).

3. House design

If you live in a tent, a yurt or a tipi your design problems will be minimal. If you live in anything larger you must make a plan. Building a home without plans is a fool's game. A mistake which can be corrected on paper with a few strokes of an eraser may, if incorporated into the building, take hours or even days to correct. Plan everything. Even build a model. It is worth the trouble.

Your house is an integral part of your life and must reflect your needs. Life is a wave. It rises and later it falls. The peak of the wave, as far as housing is concerned, arrives when the family reaches its maximum size. This is a time of stress, particularly when the children reach their teens. How severe the stress is depends in part on the building in which the family is housed. In every human being there are two impulses: *sociophilic,* tending to draw people together, and *sociophobic,* tending to pull them apart. People vary in this respect. Some are so sociophilic they can hardly bear to be left alone for a minute, some are so sociophobic that they prefer to live the life of a recluse. The majority of people live between these extremes. Sometimes they need company, at other times solitude.

Housing must be planned with these facts in mind. The true Whole-earther will limit himself to two children (zero population increase) or he will not qualify as a Whole-earther. He will plan to give those two children the privacy they need for normal development. Each child needs a room of his or her own. The house plans, therefore, should allow for the addition of two rooms to the original house if the family is to enjoy a reasonably harmonious existence.

There are only a limited number of ways in which a house can be planned. For the biological unit (male–female–two children) a house with three bedrooms, living room, bath, kitchen is optimal. If the Whole-earther cannot afford such abundance of living space and the family must make do in a one-room cabin its survival without undue strife will depend on the balance between sociophobia and sociophilia. The mating couple will do well to postpone breeding until housing is adequate. A simple cabin will provide adequate living space for a childless pair still in the early love phase, but if they wish to keep in harmony they will do well to add a room when they add a child. Later, as the children grow up and leave home the nest empties, the wave of life descends, a smaller dwelling can be sought.

Choice of house design depends largely on climate. A box-shaped, heavily insulated house or a house dug into a hillside may be just what is needed for the cold winter areas, but is inappropriate either in the humid southeast or in the arid southwest. The factors involved in planning to suit the climate have been discussed at length by Ken Kern in *The Owner-built Home*. Putting it briefly, there are two problems to consider: how to keep warm, and how to keep cool. A dweller in Florida faces a different problem from a dweller in North Dakota and needs a different house.

For warmth, get close to Mother Earth. This should be the slogan of all those who want to be comfortable without spending a fortune on heating or air conditioning. It works the other way too, Mother Earth can keep you cool. Consider, for instance, the house design offered by that true Whole-earther, Wendell Thomas (see *The Mother Earth News*, no. 10, p. 76). He buried his 32×24-foot house in the earth almost up to the roof on the north and west and up to the windowsills on the south and east sides (fig. 19, III). He called it Sunnycave and it did indeed make

140 SHELTER

use of the cave principle without subjecting its inhabitants to the inconveniences which a real cave dwelling imposes. The secret of Sunnycave (the temperature of which ranged from 60° F on the coldest winter morning to 75° F on the hottest summer afternoon) was the constant circulation of air between basement and living space. The simple shed-type roof with its 1-to-8 pitch helped to promote interior air circulation. The roof overhang kept the sunlight from the main windows in June and early July. A trellis of

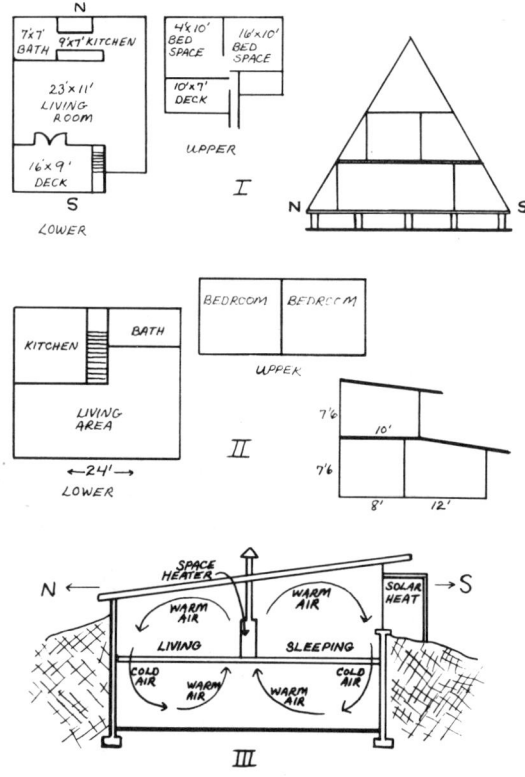

HOUSE DESIGNS, SOLAR HOUSE

FIGURE 19

grape vines kept it out through August and September. The windows were insulated at night during cold weather and those left uncovered for light were three panes thick.

There are only a limited number of ways of arranging rooms in a house. First, consider the one-room cabin. This is the simplest form of shelter. How small can one make it and still have room to turn around? What furnishings are necessary? If you wish to make do with the minimum of material, have a bed for self and mate, a table and a stove, you will need a 10 × 12-foot structure. This is the smallest log cabin described by Calvin Rustrum in his book *The Wilderness Cabin* (Macmillan Co., New York, N.Y., 1970), which offers many excellent designs and useful tips on building. By extending the roof of this structure a 4 × 10-foot porch can be created, handy for storing firewood, observing the weather, etc. This is a cold-climate cabin designed to face south with no openings on the north side.

The 10 × 12-foot living room can be rotated so that the 12-foot side faces south and can be extended on the north side to provide a bedroom (8 × 12 feet), a bathroom (8 × 6 feet) and a kitchen alcove.

In cold climates with heavy winter snow the A frame with sleeping quarters above the living quarters offers advantages. A good floor plan (taken from *The Wilderness Cabin*) is shown in fig. 19, II. This is a snug, strong design though somewhat wasteful of space. For a more economical plan using the two-story pattern, see fig. 19, I. The almost flat roof makes this house unsuitable for heavy-snow country.

Various designs will make possible the addition of rooms as the family expands. Which design is chosen depends entirely on the climate in which the house is built. In a mild climate such as that of California there is no objection to spreading your house over the earth's surface. A plan can be chosen which starts with the nuclear house (living room–bedroom, kitchen space, bathroom) and then adds rooms on either side around a patio. This gets children and adults out of each other's way and blends the house with the garden.

In very cold climates such a sprawling plan would not be practical. A cold-climate house must strive to keep its surface as small as possible and to take advantage of natural protection afforded by slopes. Wherever possible the cold-climate house should burrow part way into a hillside. Additions to such a house should take the form of an upper story rather than an extension on the surface. The solution is to use a roof sufficiently steep to allow for the construction of two additional bedrooms. The upstairs rooms can be rendered much more comfortable if arrangements are made when framing the roof for the addition of dormer windows. Such windows make possible cross ventilation, a highly desirable feature during the height of summer.

The true Whole-earther who designs his own home will be wise to bear in mind that he may one day wish to sell it. For this reason he may prefer to avoid freakish designs. Although he himself may feel perfectly at home living in a truncated dodecahedron, others may not. Any half-way effective house will be hard work to build and it is a pity to put that hard work into a structure that cannot be sold at least for what it cost to build.

For this reason I have not paid much attention to the current passion for domes. Theoretically a dome is a noble structure, capable of enclosing almost unlimited areas. Practically, for the do-it-yourself builder, the dome is a source of endless headaches and the result is rarely worth the trouble. The angles are complex and errors are hard to avoid. Wood, unless carefully seasoned, tends to warp and the dome tends to leak. For people who love togetherness the dome's single space may be fine but for sociophobes who need privacy dome living is as bad as living in a goldfish bowl. Finally, the structure is hard to sell should you ever wish to move.

4. Choosing the site

There are five points to be borne in mind when choosing the site of a house.

(1) Exposure. The natural place for a house in a cold climate is a south slope protected on the north by trees. Trees are desir-

able to provide shade and a windbreak and should be planted if they are not already there. Avoid planting trees so close to the house that they will constitute a menace when they grow large. In a hot climate the house should be oriented to catch the prevailing breeze.

(2) Water supply. Always try to locate your house in such a way that water, whether piped in from a spring or drawn from a well, can be brought directly into the house and preferably stored in an overhead tank for ease of distribution.

(3) Drainage. Avoid locating a house in a hollow or on very heavy clay. If you are building near a river, make inquiries as to the possibility of floods.

(4) Proximity to road. The true Whole-earther who locates in the wilderness may want to get as far from roads as possible. Less resolute souls may be more concerned with getting children onto the school bus, getting out to a job or for shopping. Roads, though sources of pollution, noise and accidents are also life lines. If you build close to a road, try to locate your house above it rather than below it.

(5) Zoning regulations. These should be checked before you choose a site. They specify how close you can build to your neighbor's property.

5. Laying foundations

Whether you build a log cabin or a frame house you can be sure that the house will be only as good as its foundations. Furthermore, if you want a rectangular house, that foundation must be rectangular. A serious error in the foundation will plague the builder from beginning to end of the work. It is therefore essential, having selected a site with due regard for water supply, accessibility, sun, shade and local ordinances regarding proximity to your neighbor, that the builder lay out his building with caution and awareness. The best way to avoid mistakes is to buy plenty of builder's nylon twine, 2×4's and planks for the making of batter boards. A 50- or 100-foot steel tape is also of value. Decide which way you want your building to face and drive two stakes into the ground, marking the front wall. Measure off the width of the house and drive in two more stakes. Erect batter boards, as shown in fig. 20, 3 or 4 feet back from the stakes at each corner. The batter boards must be at the same elevation, so check this by means of a hose (fig. 20). Weighted lines can be

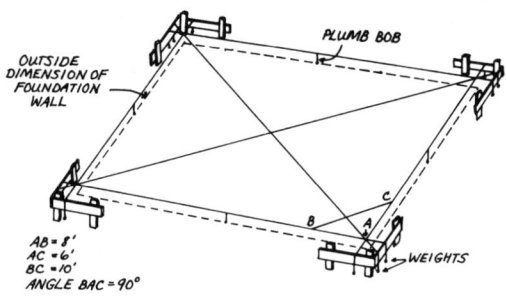

LAYING FOUNDATIONS

FIGURE 20

hung over the batter boards and a plumb line dropped from the intersection to a nail driven to the top of a stake. Next adjust the lines in such a way that a point 6 feet from A on line AB and a point 8 feet from A on line AD are exactly 10 feet apart. The angle DAB will then be a right angle. The other strings are adjusted in such a way that the four stakes are exactly the right distance apart. A second check can be made by measuring diagonals AC and DB. If these are exactly equal you have a rectangle. Once the lines are laid out, the position of the foundation trenches can be indicated by taking a handful of slaked lime and running it along beneath the string.

6. Building materials

The cheapest, most readily available building material is earth. And the most down-to-earth of all dwellings is the cave, natural or man-made. There are some kinds of rock (we have them here in Sonoma County, Calif.) that can be cut almost as easily as cheese but which harden when exposed to air. They may be volcanic tufa or certain forms of limestone. To cut one's way into such rock with a sharp mattock or adze may be a rewarding experience in itself. The site selected for such burrowing should be a south slope and the excavated material can be piled outside the entrance to make a terrace. The cave should be carved with an arched roof for strength and supporting pillars left to prevent fall-ins. This type of burrow can be constructed on very steep slopes which are otherwise worthless for building.

The cave dwelling has a number of obvious drawbacks. Though cool in summer and warm in winter owing to its natural insulation, it is almost certain to be damp. Dwellers in burrows, unless very sure of the stability of the rock in which they build, may be troubled by the thought that a cave-in may convert their house into their tomb. Heating may be difficult (where does one put the chimney?) and plumbing may also present problems. You may also find yourself persecuted by the local bureaucrats, who may regard cave dwellings with disfavor.

Many of the drawbacks of the cave are eliminated in the semi-cave. This can be built on almost any steep slope, provided one is willing to put in retaining walls. Like the cave it is dug into the side of the hill but instead of relying on a roof of rock the builder roofs over the excavation. The sides and back are of earth or rock. The front is a wall built of wood or stone, with doors and windows. Illumination, which is such a problem in the cave, is obtained by putting skylights in the roof. The wide overhang drains the rain away from the walls and prevents the place from becoming too damp. If caving-in of earth walls is a problem, face the earth with a second wall of wood or stone and include a layer of waterproof material between earth and inner wall.

Next to building in the earth the true Whole-earther may favor building a house of earth. There are four ways of building with earth: wattle-and-daub, sod, adobe brick and rammed earth. Wattle-and-daub is fastest and probably easiest. You put up

146 SHELTER

forked corner posts wherever you want corners. (Though four corners are traditional no one need be bound by this. Hexagons, heptagons, octagons, nonagons are all possible.) Between the forked corner posts you place a sill of some rot-resistant wood, drill holes at 6-inch intervals, drive stakes of bamboo, willow or hazel into the holes. Or simply drive the stakes straight into the ground. With willow wands, bamboos or bundles of grass, make a wattle by weaving this filler material between the stakes (fig. 21). When you reach the height of the corner posts you stop

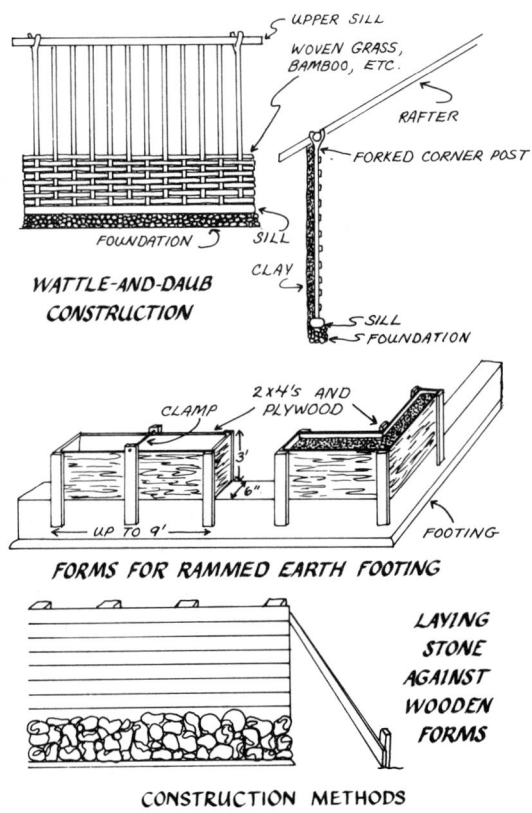

FIGURE 21

weaving and put crosspieces in place to support the roof members. Insert bundles of dry grass or straw into the wattle to help bind the daub. Make a mud of clay and some sand (too much clay cracks, too much sand won't hold together). Smear it all over the wattle with a board or mason's trowel. Allow to dry, add a second layer and a third. If you can finish with a layer of cement it will add greatly to the strength of the wall. The inside of the house should also be daubed and both outside and inside whitewashed. The roof can be built of wooden poles held up by a center post or, if the house is rectangular, meetings at a ridge pole. Ordinary straw thatch or a layer of wood shingles will serve to keep out rain, but straw thatch, if it is to be effective, must be about 2 feet thick. The roof overhang on a wattle-and-daub house must be sufficient to prevent rain splash from eroding the base of the wall.

A sod house can be constructed only where there is plenty of sod. The sod is formed by the grass roots and dead stems that make a tangled mass in the upper 8 inches of soil, which can be cut with a spade and shaped into bricks. Sod houses were widely used in the early days on the prairies because sod was the only building material available. It does make a fairly warm and sturdy shelter, provided the walls are at least 2 feet thick. The bearing members for the roof are best made from forked posts, or, if the roof rests on the sod walls, sills of squared wood must be placed on top of the walls to distribute the weight.

Rammed earth is a building method, used particularly in North Africa, to construct buildings of considerable size and charm. The technique depends on making wooden forms to hold the earth, then filling the forms with damp earth and ramming it until it is compact. The forms can be made of 1×6 boards nailed to 2×4's. They can be about 9 feet long, 3 feet deep and 18 inches wide, held together by metal spacing pieces. The corner forms can be constructed as shown in fig. 21. Rammed earth walls are only as stable as the footings on which they are built and these footings in cold climates must extend below frost line (3, 4 or even 6 feet). The cheapest form of footing is a trench 2 feet wide filled with rocks and cemented on top. Built on such a footing a rammed earth wall will be stable. Frames for doors must be put in place and braced before the walls are started. Frames for windows must be placed in position as the walls rise and should reach to the top of the wall so that they will not need to bear a load of earth as well as the weight of the roof.

Building with adobe brick is probably more time-consuming than building with rammed earth. It has two advantages: (1) If you are not in a hurry you can make your bricks the summer before you start building. (2) Once the bricks are made the walls go up fast. Adobe bricks can be made from any earth that has enough clay to hold it together and enough sand to prevent cracking. A good mix is 50% clay, 30% sand with 15 to 18% moisture. The soil can be mixed with chopped straw, about 150 pounds of straw per 1,000 blocks. The earth-and-straw mix should be puddled in wooden forms $4 \times 12 \times 18$ inches. A form to make four bricks at a time is convenient to handle. The bricks shrink as they dry so that the form can be easily removed. Adobe bricks can be waterproofed with Bitumil (a Standard Oil product). Or the adobe wall can receive a coat of stucco, which will both waterproof and strengthen the wall.

No matter how you do it, building in earth is a sweat. Edward Allen, author of *Stone Shelters* (MIT Press, 50 Ames St., Room 765, Cambridge, Mass. 02142), sums up the matter in the *Last Whole Earth Catalog* (p. 102): "It's more work than with any other material, bar none. It's slow, heavy and sweaty. You move tons of the stuff every day to make a few square feet of wall . . . Earth is great in warm climates but in any moderate or cold climate it'll give you rheumatism by Thanksgiving time."

So much for earth. Stone is better. If you have plenty of field stone, plenty of patience and a good supply of cement a very good-looking shelter can be built. But it does take skill and time. Working with untrimmed field stone is an art that can be learned only by practice. A stone wall, like an earth wall, must have massive footings. Mark Mendel of Dixmont, Me., writing to the *Last Whole Earth Catalog* (p. 103), describes a stone shelter he built with foundations 4 feet deep on a 1-foot poured-concrete footing and the wall was 1½ feet thick up to 1 foot above the ground. This is what you face if you live in Maine. By far the easiest way to build in stone is to use wooden forms, properly braced and leveled, and use cement against the form and stone on the outside (fig. 21). The material in the forms can later be used for floors or the roof. This form of building gives you a smooth interior finish and an exterior of stone. It was used by Helen and Scott Nearing in their house in Vermont (see *Living the Good Life*).

7. Log cabins

The log cabin is a form of housing suitable for areas having abundance of trees and lack of sawmills. It is not an economical way of using wood for a building. A better house can be constructed more quickly and with far less wood if the trees are cut to dimension lumber and used to construct a frame house. This, however, calls for a saw large enough to transform tree trunks into 2 × 4's, 2 × 6's and planks. Before the days of sawmills all such sawing was done with pit saws, which might be up to 12 feet in length. Two men were needed. One stood in the pit, one stood above. The pit saw was in fact a very large rip saw and cut only on the downstroke. Such saws can still be obtained and any Whole-earther who wants to enjoy some really strenuous labor should learn how to use one.

The modern version of the pit saw is the chain saw. Chain saws are dangerous, smelly and the noise they make is deafening but if one can stand their racket and afford their price they do represent the nearest approach to a one-man sawmill. Anyone wishing to build a log cabin can simplify the work if his equipment includes a chain saw.

A log cabin, like any other building, will be only as strong as its foundation. The trench should go below frost line and be extended above the soil to about 18 inches. There is a belief, mentioned in *The Foxfire Book* (Eliot Wigginton, Doubleday, New York, 1972), that termites cannot climb higher than 18 inches. The belief is probably not correct but it is always a good idea to get your wooden house well clear of the ground. You can use rock, poured cement or concrete blocks for the foundation. A continuous foundation is certainly the best as it provides insulation, a crawl space under the building and a strong support for the walls. Sufficient vents should be introduced into the foundation to ensure circulation of air.

It may happen, however, that you are not able to make a continuous foundation because you do not have either the time or the materials. In this case it is possible to build your house on piers either of wood or concrete. Excellent piers of wood can be made from old telephone poles. The wood should be thoroughly creosoted and dug into the ground below frost line or sunk to bedrock. Allow the wooden pier to project above ground somewhat more than will be necessary. If you are building on a slope, trim the pier that is closest to the ground to give at least 1-foot clearance. The set-up shown in fig. 22 makes cutting easier. Carefully level all the other piers from pier I by the hose method (fig. 20), then cut them. The wooden piers should be spaced 6 feet apart in rows 6 to 8 feet apart.

Concrete or masonry piers are more satisfactory than wooden ones because they do not rot. After digging a hole down to bedrock or below frost line, place in the hole a wooden form 10 × 12 inches. The form should project at least 6 inches above ground. A block of wood sunk in the concrete will provide a nailer for a wooden post which will rest on the concrete pier and support the floor. This combination of wood and concrete is desirable if the ground slopes and the distance from ground to floor is more than 18 inches. If the ground does not slope much the concrete forms can be carefully leveled and filled to the top. A ⅜-inch anchor bolt should be bedded in the concrete and project enough to go through the log sill and leave enough thread to take a washer and nut.

If you are not in a hurry it is an excellent idea to give your foundation a whole winter to harden and settle before adding the superstructure. You will, in any case, be well advised to give your logs time to season before you use them. The ideal order of operations is to put in the foundation in early fall before frost hardens the ground, cut the logs in late fall, peel off the bark with an axe or drawknife, stack the logs on leveled skids. If you have more than one tier of logs, place peeled poles between the tiers. Logs with slight curves should be piled so that the arc is up and the weight of the other logs will help to straighten them. Logs between 9- and 12-inch diameter are best. If the logs are to project beyond the cabin corners they will need to be cut longer than the outside of the building.

Building begins with placing the sills on the foundations. The

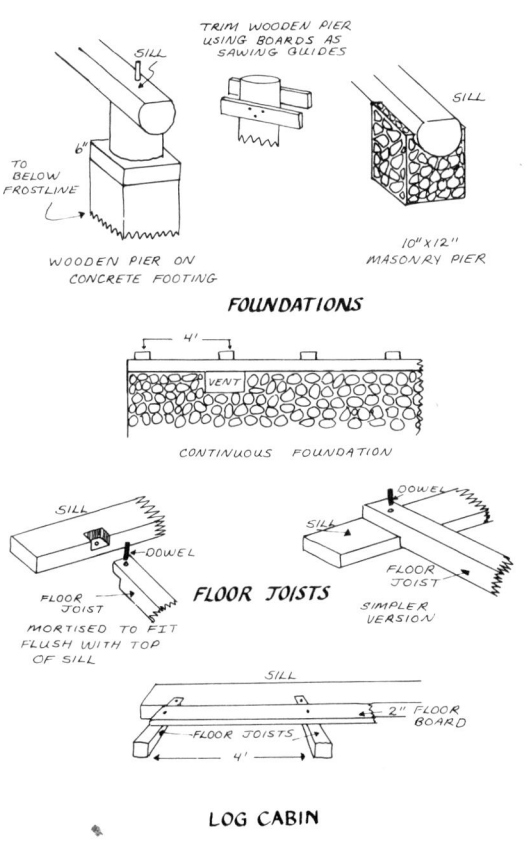

LOG CABIN

FIGURE 22

sills take the weight of the house. They must be straight, between 8 and 12 inches thick and smoothed top and bottom with an axe or chain saw. They must also be of the same thickness. The sills are bolted to the foundations. If the cabin is 20 feet wide a central sill must be placed in the middle to give extra support to the floor joists.

The joists which hold up the floor span the distance between the sills. They should be placed on 4-foot centers. If you want an authentic log cabin you will lap-join the floor joists into the sill and secure them with a dowel. The floor joists must be smoothed with an axe, adze or chain saw on the top side and they must be flush with the sill. In *The Foxfire Book* the floor joists are called sleepers and an alternate method of joining them to the sill is shown in fig. 22. The floor joists must be notched to fit the central sill.

The flooring for the old-fashioned log cabin was split out in the form of puncheons 4 feet long which were trimmed smooth with an adze and attached to the floor joists with dowels. Splitting such puncheons calls for skillful use of the froe, an instrument also used for splitting shingles. An easier method of cutting floorboards is to use the chain saw with a suitable guide to keep it level. If your floor joists are on 4-foot centers your floorboards should be at least 2 inches thick to give the necessary rigidity to the floor. They can be secured either with 16-penny nails or with dowels.

Once the floor is in place work can begin on the walls. If your logs are in short supply and if you have a chain saw you can economize in logs by cutting them down the middle and using half logs for your cabin walls. The logs should first be trued with the chain saw so that the edges fit together snugly. This is done by clipping the logs together with two iron log dogs and running the blade of the chain saw between the logs. If they still do not fit the logs can be brought together again and the operation repeated. Then anchor the log with the trimmed side uppermost and cut it straight down the middle.

The half-log method lends itself admirably to construction of the stockade-type cabin with the logs placed vertically instead of horizontally. Though the horizontal logs are traditional and more picturesque the vertical logs offer several advantages. They are relatively short (about 8 feet) and therefore easy to handle. They offer no horizontal crevices through which rain can enter. They do not have to be elaborately notched at the corners but can be secured by nails or dowels to the sill and top plate. They leave a flat surface on the inside to which some sort of insulating wallboard can be attached.

The stockade-type log cabin must, of course, have a frame consisting of uprights and a top plate to which the logs can be attached. Such a frame can best be made from 4 × 4 squared lumber. The corner posts should be mortised into the sill. The plate can be secured to the posts by dowels and glue. Diagonal bracing at the corners will add to the strength of the frame. The double half-log method described by Calvin Rustrum (*The Wilderness Cabin*, p. 92) gives a strong, well-insulated wall of the stockade type. Half logs can also be trued with a chain saw and used horizontally notched as shown in fig. 23.

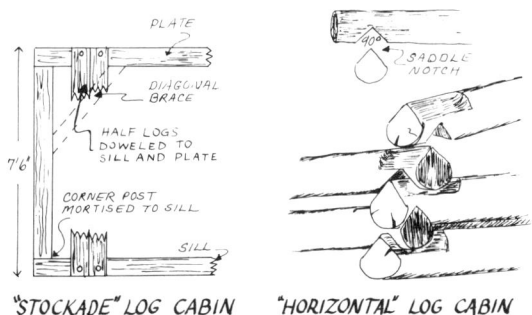

FIGURE 23

The true Whole-earther who regards the chain saw as an invention of the devil will, of course, have no part in such methods as these. Confining himself to hand tools, a hand saw, axe, adze, froe and auger, he will "notch up" his cabin in the traditional manner. This notching, beautifully illustrated in *The Foxfire Book*, can be a work of art on the level of cabinet making. The most elegant notch is the dovetail. This takes real skill and lots of time and requires trimmed logs. The saddle notch is easier (fig. 23). The peak and the V must both be right angles or the fit will not be perfect. To hew these notches with a hand axe requires practice, also very sharp tools. With this method the ends of the logs can be made to project 6 inches beyond the corner of the building, giving the traditional log cabin appearance.

When the log wall has been taken to the required height the top log must be carefully leveled to form a plate, if this plate is not already in place as in the stockade-type structure. The plate will carry the ceiling joists, which not only support the ceiling but also prevent the walls from being pushed out by the pressure of the roof. These joists must be cut to fit the top plate and pegged in place with dowels. The walls can then be carried up a further 4 feet to give a loft handy for storage or the roof can be put on directly. If your cabin is more than 20 feet across, you will have to provide a central bearing wall to support the ceiling joists.

Once the walls are up, it is necessary to choose straight logs for rafters. An authentic log cabin roofed with split shingles must have a pitch of 45° to make sure that water runs off the roof rather than through it. The 90° saddle notch can be used to secure the rafters to the top plate but the angle must be carefully adjusted to allow for the 45° angle of the rafter. The rafters meet at the ridge pole, to which they can be secured by dowels. "Wind beams" are sometimes pegged to or mortised into the rafters about 4 feet below the peak for added strength. Vertical supports resting on the ceiling joists may also be added.

To close the gable end, either run vertical trimmed beams up to the end rafters and dowel half-round logs to them or go on building the log wall up to the peak. As these gable-end logs cannot be notched they must be doweled together.

The roof itself is traditionally built of purlins (laths, as they are called in Appalachia) to which roofing shingles are attached.

HOUSING 155

If they could get them the old settlers used 2-inch-thick boards, which they nailed to the rafters. If you cannot get boards poles will do.

You can make your own shingles from cedar or oak by using a poleaxe, a go-devil, iron and wooden wedges, a froe and a mallet. For roofing shingles a section of tree trunk 2 feet long is split into bolts, the heartwood is split out, the bolts are debarked and rived with a froe. To use a froe effectively one must make a board brake (fig. 24). This can be made from a Y crotch of any

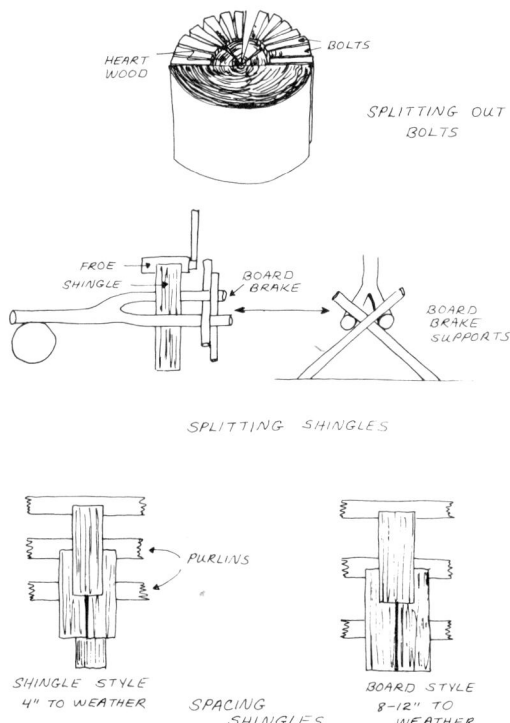

MAKING AND LAYING SHINGLES

FIGURE 24

hardwood, supported on a log at one end, held up by two poles at the other. The froe is hammered into the bolt with the maul. The bolt is placed in the board brake and carefully worked in such a way that it splits off a shingle. The crack in the wood tends to run out on top and can be corrected by pressing down on the bottom half of the bolt. Splitting out shingles with a froe is an art that can be learned only with practice.

Shingles, says *The Foxfire Book,* can be laid in two ways, shingle style with 4 inches exposed to the weather, or board style with 8–12 inches exposed. Which style you use will depend on the pitch of your roof. On a steep roof the board style may be adequate but on a 45° roof it is safer to use the shingle style. The spacing of the purlins will also depend on which style you use. It is advisable to extend the purlins about 2 feet beyond the gable ends for added protection from rain.

The top two rows of shingles should be laid over metal flashing or sections of heavy tarpaper. It is difficult otherwise to prevent leaks at the ridge. If your cabin is so designed that it has a valley in the roof you will have to flash this also.

8. The frame house

A frame house is easier to build than a log cabin and uses less wood. For this reason a true Whole-earther may find it worthwhile to set up his own sawmill and reduce the logs that would otherwise form a log cabin to standard-size lumber. Provided you have a fairly large chain saw you can create an effective sawmill simply by placing logs in a suitable frame of angle iron with guides and moving the saw along in a horizontal position. It helps to have someone to guide the other end of the blade. The lumber so produced will be rough but adequate.

Whether he cuts his own dimension lumber or buys it readycut the house builder must know what sizes of wood he needs. It is true, as Ken Kern points out, that the traditional frame house is overbuilt, that vertical 2×4 studs on 16-inch centers represent a waste of wood and add a fire hazard. The builder, however, must be sure he knows what he is doing before he makes major changes in the framing of his house. It is better to overbuild than to have the structure collapse. Also, if he is building under the eyes of the bureaucrats he may find he has no choice but to follow the code.

A frame house consists of four elements: foundation, floor, walls and roof.

(1) Foundation. The best foundation is poured concrete with an 8-inch footing for a one-story home, a 12-inch footing for a two-story home. The footings must go to bedrock or below frost line. The foundation walls should project 6 inches above grade. The loose rock foundation is satisfactory, provided the trench goes below frost line. In this case, cover the top layer of rock with 6 inches of cement and extend the walls with cinder block or cement block. Piers of wood or masonry dug down to bedrock or below frost line will make a cheap though not a good foundation. A full basement with a 7-foot ceiling is a tremendous advantage in cold climates. The basement walls must be built on footings of poured concrete 8 to 12 inches wide and 6 inches deep with a reinforcement rod (#4 rebar) embedded in the concrete. The basement walls can be cinder block or cement block. The outside of the walls should be stuccoed and waterproof. If you do not want your basement to turn into a swimming pool in wet weather, surround it on the outside with a drain to take off the water. If you have no slope where you build you must put in a sump and pump out the water. In the absence of a full basement a 4-foot crawl space will help keep the house warm in winter, cool in summer.

You can do away with basement and crawl space and build your house on a concrete slab. The method has little to recommend it. It involves tons of concrete, large amounts of welded reinforcing mesh and rebars. If you do use this method, be sure to run all plumbing into the house and get it accurately positioned *before* the slab is poured. You will never be able to get at it again.

(2) Floors. The floor has four parts: sill, floor joists, subfloor, top floor (fig. 25). The sill rests on the foundation. If the founda-

158 SHELTER

tion is continuous the sill can be a 2 × 6 anchored to the foundation by means of anchor bolts (½ × 10 inches, 72 inches on center). If you use posts or piers 6 to 8 feet apart you will need heavier sills (4 × 10-inch or 4 × 12-inch). After anchoring the sill, lay out the floor joists.

The distance apart of these joists and the size of lumber required depends on whether you use post-and-beam-type construction or follow the more conventional framing plan. The post-and-beam-type structure is popular in California and accepted by the building inspectors. It uses heavier lumber but less of it. A floor built by conventional methods having a span of 15 feet

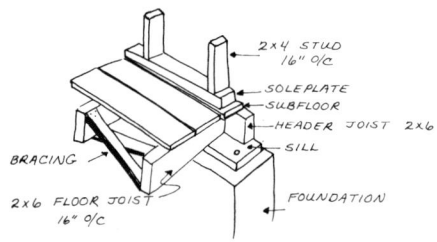

FLOOR AND WALL CONVENTIONAL FRAMING

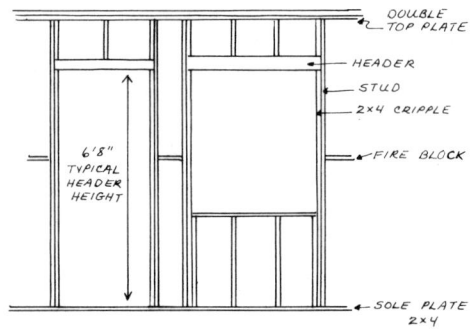

FRAMING FOR DOORS AND WINDOWS

FIGURE 25

would need 2 × 10-inchers on 16-inch centers. By the post-and-beam method one would use 4 × 12-inchers on 48-inch centers. The size of the joists can be greatly reduced if they are supported on 4 × 4 posts placed on 16 × 16-inch concrete footings at 6-foot intervals. The relationship between size of floor joist and span is shown in table II for Group I lumber (Douglas fir, construction grade). Hemlock, fir and spruce are weaker woods and the spans for which they can be used are about 1 foot less than Douglas fir.

TABLE II
Allowable spans for floor joists and roof rafters not supporting ceilings

Size, inches	Spacing, inches on centers	Floor joists, feet	Ceiling rafters, feet	
			Slope 1:3 or greater	Slope less than 1:3
2 × 6	16	10	15½	14½
2 × 8	16	13½	20	19
2 × 10	16	16½	25½	23½
2 × 12	16	20	—	28
4 × 6	48	8	12½	12
4 × 8	48	10	17	15½
4 × 10	48	13	21	19½
4 × 12	48	15	—	23½

If you use conventional framing you must (a) nail your joists (2 × 6, 2 × 8 or 2 × 10) on 16-inch centers to the sill, (b) secure the ends of the joists to the header joist with two 16-penny nails, (c) install solid blocks or diagonal bridging every 8 feet to keep joists rigid, (d) nail subfloor of 1 × 6 or 1 × 8 tongue-and-groove to every joist, and later, when the house is closed in, (e) cover the subfloor with building paper, (f) lay the finished floor on the subfloor. Post-and-beam construction is simpler. The heavy beams (4 × 8, 4 × 10, 4 × 12-inchers) are laid out on 4-foot centers and nailed to the sill. They require no bridging. By using 2 × 8 tongue-and-groove planed on one side the need for a top floor is eliminated. Or utility grade 2 × 8 tongue-and-groove can be used and covered with light plywood or fiberboard over which can be laid a carpet.

(3) Walls. Walls in a conventional frame house have three parts. The *sole plate* is normally a 2 × 4 that forms the base of the wall. To it are nailed the *studs,* also normally 2 × 4's on 16-inch centers. To the tops of studs is nailed the 2 × 4 top plate and a second 2 × 4 is nailed to the first to give added rigidity. Once the subfloor of a frame house is finished, the walls can be assembled quickly and easily on the deck, raised into position and braced with diagonal bracing. It is necessary to use special corner posts arranged to give a surface to which wallboard can be nailed. Openings for doors and windows always take the same form (fig. 25). The *header* is the piece going across the top of the opening. As it must bear the weight of the roof or second story if there is one it has to be strong. For a 4-foot span, use a 4 × 4, up to 6 feet a 4 × 6, up to 8 feet a 4 × 8. The fire blocks are bits of 2 × 4 nailed between the studs. A diagonal brace (a 1 × 6-inch board) let into the studs gives added rigidity to the wall but is not required if 4 × 8-foot sheets of insulation sheathing or plywood are used on the outside wall.

It is possible, and permitted by code in California, to use 4 × 4 posts on 48-inch centers in place of conventional studs on 16-inch centers because the 4 × 8-foot sheets of wallboard can be nailed to the studs. In cold climates *insulation sheathing* is generally nailed to the outside of the studs and the exterior siding is nailed over this sheathing. In California 4 × 8-foot sheets of exterior plywood are popular. They are nailed directly to the studs and give such rigidity to the walls that diagonal bracing is not necessary. On the inside, walls can be finished with plywood or sheetrock. Electric wiring and plumbing must all be installed before the walls are closed in.

The height of a wall from floor to ceiling is generally 8 feet. A minimum ceiling height of 7 feet 6 inches is allowed by code, with 7 feet permitted in hallways and corridors.

(4) Roof. The structure of the roof depends on the climate in

which the house is situated. For climates in which heavy snow is not expected a shed roof of heavy beams is by far the easiest to construct. The beams are placed on 4-foot centers and should lie directly over the studs. The size of beam will depend on the span (table II). The slope of a beam-and-plank roof can be as little as ¼ inch in a foot. The planks (2 × 8 tongue-and-groove planed on both sides and kiln-dried) can be nailed directly to the beams and left exposed. Over the planks a layer of 1-inch insulation should be nailed. This almost flat roof has to be waterproofed with three layers of tarpaper bonded together with hot tar and finished off with gravel spread over the last layer of tar while it is still hot. Roll roofing can also be used but must be well lapped and properly glued at the edges or leaks are inevitable.

In order to use shingles on a shed-type roof the true Wholeearther will have to increase the pitch to 1 in 3. This can be done without notching the roof beams if advantage is taken of metal holders which can be either purchased or made from galvanized sheet metal. The holders are nailed to the top plate and the beams are nailed in the holders. If you use beams on 4-foot centers you will have to use 2 × 8 tongue-and-groove to sheath the roof. If you use lighter rafters and split shakes or shingles you can economize in materials by using 1 × 4-inch or 1 × 6-inch purlins with 4-inch spaces between them. The first three rows of purlins at the eaves should be solid.

If you want to be sure of a tight roof, buy some rolls of 18-inch-wide #30 building felt and introduce it between each course of shakes. Start laying shakes at the eaves. Put two nails in each shake and lay a *second course of shakes* on top of the first with the top shake covering the gap between the two shakes below. Then lay the next course. The exposure to weather for hand-split shakes varies with shake length as follows: 18-inch, expose 8½ inches; 24, expose 10; 32, expose 13. Nail the shakes in place at opposite ends of the roof, then run a chalk line across from the lower corner of one to the lower corner of the other and snap the line. This will enable you to get your rows of shakes straight. Commercially sawn red cedar shingles are easier to lay than shakes and don't require interleaved building felt. They are expensive. Asphalt shingles are still easier to lay, cheaper than red cedar and more fire-resistant but for these you must sheath your roof solid and lay building paper over the sheathing.

A shed roof with a 1:3 slope will result, if you leave it exposed, in a high ceiling on one side of the house. This is fine in a mild climate, not so good in a cold one. If you want to add a ceiling for extra warmth this can be done by framing the house as shown in fig. 26. The ceiling joists should be on 16-inch centers and insulation should be placed between them.

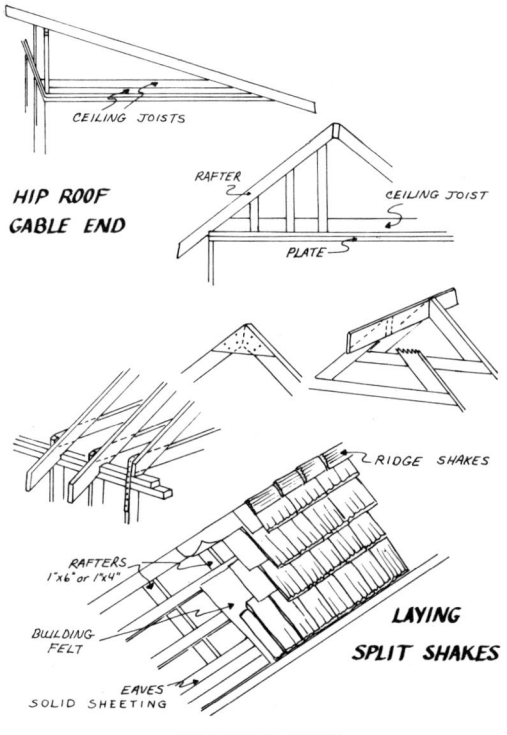

FIGURE 26

If you install a ceiling you lose most of the advantages of a shed roof and might as well build a hip roof or gambrel roof. The hip roof is commonly constructed by notching the rafters to fit the plate and nailing them to a ridge rafter at the top (fig. 26). Ceiling joists are nailed to the rafters as shown. The rafters are toe-nailed to the plate. If you want to avoid the ridge board a plywood gusset can be used to hold the rafters at the peak. The join will be enormously strengthened if you glue as well as nail these gussets. A rafter tie (1×6 or 1×8-inch) is desirable to give added strength. A handy trick for keeping your roof from blowing off in very windy sites is to use metal straps nailed to the studs and ceiling joists.

Once the rafters are up, the floor can be closed in either with plywood sheets or 1×8 tongue-and-groove. Rafters can be spaced 24 inches on centers but in heavy-snow country it is wiser to follow the conventional practice of placing one rafter on 16-inch centers over each wall stud. The ceiling rafters will also be spaced 16 inches on centers.

If you want more room upstairs without adding a second story a gambrel roof is worth considering. This is the kind of roof you often see on barns. It has two parts, a steep section rising to 7 feet 6 inches above the top of the ceiling joists and a less steep section to the ridge. It is a fine roof for snow country. If you frame openings for dormers you can add these later (fig. 27). The opening consists of double rafters and double headers with temporary pieces in between to be removed later. For dimension of roof rafters, see table II. Remember, if you contemplate putting rooms upstairs, that your ceiling joists will have to double as floor joists. For this reason you should estimate the dimension of such ceiling joists from the section on floor joists in table II.

The thatched roof, though very common in England, seems to be practically unknown in America. This is a pity as a well-thatched roof is warm, waterproof and quite durable. A roof thatched with wheat straw has a lifetime of about 50 years. One thatched with reed (*Phragmites communis*) will last for a hundred.

Any Whole-earther who chooses to thatch his own house will

have to prepare his own materials. Wheat straw as it comes in the bale is completely useless. To get straw suitable for thatching one must cut the wheat with a scythe or sickle, knock out the grain with a flail or stomp it out with one's feet, bind the straw into bundles.

However, if the Whole-earther has access to a tule swamp he can collect reed, in some cases mixed with cattails, which makes the best thatch. Reed should be tied in bundles about 12 inches in diameter. The bundles of thatch are tied to the purlins, starting

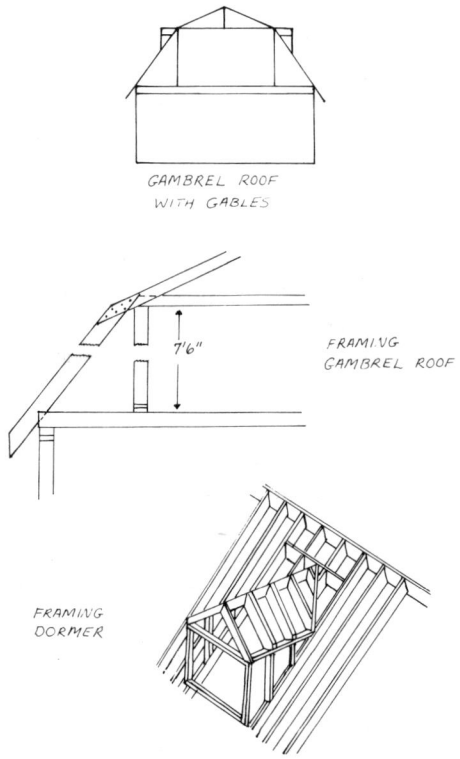

FRAMING GAMBREL ROOF AND DORMER

FIGURE 27

at the eaves. The roof must have a pitch of at least 45°. The bundles are laid like shingles, each one covering the gap between the two bundles below. They are beaten into their final position with a wooden leggat, a square board with a handle. At the eaves and peak of the roof the thatch is secured with thatching spars of split hazel holding down lengths of split hazel called liggers. The pattern made by the liggers is quite distinctive and professional thatchers vary the design, using it like a signature. (For a good account of the thatcher's art, see J. Arnold, *The Shell Book of Country Crafts,* John Baker, London, 1968.)

9. Heating and cooling a house

A house is a micro-environment, a little portion of the great outdoors enclosed by walls. To keep this small world comfortable for its inhabitants it is necessary to adjust the temperature.

Ken Kern in *The Owner-Built Home* has summarized this comfort problem in one sentence: "We are comfortable when we give off heat effortlessly *at the same rate* that we produce it." If the body heat escapes more rapidly than it is generated, chilling results. If the body heat cannot escape rapidly enough, overheating results. One can die of cold; one can also die of heat stroke.

The true Whole-earther in the United States is probably more apt to suffer from cold than heat. He is most likely northward bound because only in this direction can he find reasonably cheap land. House heating will therefore be considered first.

If you have followed section 3 on house design you have already simplified your heating problems. You have (a) dug into a south slope if you have one, (b) used banked-up earth as insulation on the north side, (c) dug out a 4-foot crawl space below the house and made arrangements for air to circulate between the crawl space and house, (d) arranged your south-facing windows to trap as much winter sun as possible, (e) prepared insulating screens with which to cover the windows during nights and cloudy days. All that remains to be done is to choose a heating unit that will use your fuel (probably wood which you have cut by the sweat of your brow) as effectively as possible.

If you want to live really simply the first thing to do is to use one piece of equipment for both heating the house and cooking your food. The ideal to strive for is a stove that will hold as much heat as possible. This can be done by making the structure quite massive, incorporating a hot water heater and an oven. You can do this by using a 50-gallon oil drum as a base. From various war surplus outfits you can buy very cheaply a kit for converting such a drum into a stove. But instead of leading the flue gas out of the top you lead it out of the bottom, making it circulate around the heating element of your water heater. You can also construct an oven at the rear of the drum and, by installing a sand box on top of the drum, make places for your cooking pots. If you enclose the whole thing in masonry (stone or adobe brick) you will conserve heat as well as making the structure esthetically pleasing. The masonry absorbs heat when the fire is hottest and reradiates it later. If you build a well-insulated curtained sleeping shelf above the stove you will be able to enjoy warm nights even when the temperature outside is well below zero.

It may be argued that all these effects may be obtained with an old-fashioned kitchen range. They may, but the kitchen range does not have the heat storage capacity of this massive masonry structure, is a pretty weighty item to transport back into the woods and costs new from $200 to $600 (Portland Stove Foundry, Sales Office, Box 59, Sterling Junction, Mass. 01565).

The enterprising Whole-earther can build this space heating–food cooking unit himself without relying on technology's dubious gifts (50-gallon drums) or the by-products of America's war machine. It can all be built from stone or adobe brick and lined with clay. If you want to introduce the oven and water heater you will need sheet metal, some cast-iron pipe and a water tank. It is also nice to have a fire door, otherwise the rate of burning of the fuel cannot be regulated. The structure, if built entirely from masonry, will be very heavy and should be constructed on a poured-concrete footing 12 inches deep, and 6 inches wider on all sides than the structure itself. This applies to all masonry fireplaces and chimneys. The fireplace and chimney must stand on its own footing and be independent of the house.

All that one needs to know is how to frame the openings around the chimney and how to use roof flashing to prevent leaks at the junction between roof and chimney. To frame the opening in the floor one uses a double header and nails the floor joists to it. To frame the opening in the roof one uses a similar device. Aluminum or copper flashing is used to waterproof the gap between chimney and roof (fig. 28). If your chimney emerges at the

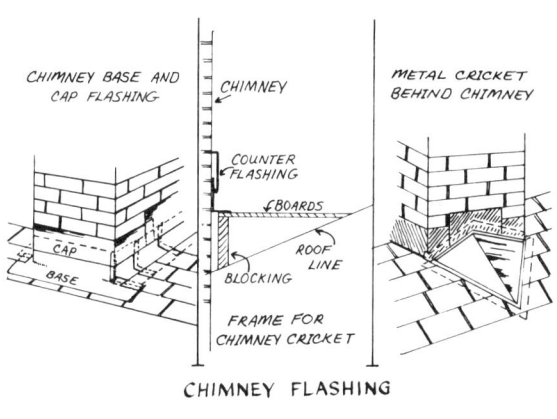

CHIMNEY FLASHING

FIGURE 28

ridge of the roof, build it 2 feet above the ridge. If it is located at the eaves, determine the height of the roof 10 feet from the chimney and build it 2 feet higher.

If you use metal in masonry when building your heating unit, remember that metal expands more than masonry when heated. This means that it will crack the masonry unless an elastic layer is introduced. This elastic layer can be loose asbestos or fiber glass. A layer of this material introduced between the metal and masonry will save much trouble later. It is also wise when building a chimney to use fire clay flue liners ($8\frac{1}{2} \times 13$ inches or 10-inch diameter if round). These liners prevent accumulations of soot and give the amateur a core around which to build the chimney.

If you must use a different source of heat for cooking and space heating the best and cheapest device is a simple wood-burning stove, with metal flue pipe which acts as an auxiliary heater. The Ashley Thermostatic Wood Burning Circulator costs $80 to $120 (from Ashley Automatic Heater Co., P.O. Box 730, Sheffield, Ala. 35660) and offers the great advantage that it controls its own rate of burning. By all means, avoid the old-fashioned open fire unless you want to spend the entire winter hauling and sawing wood. If you get a special thrill from watching flames, build the equivalent of a Franklin stove with a large surface to absorb and re-radiate the heat. The old-fashioned fireplace is a ridiculous device because it takes warm air from the room, uses it to burn the fuel, then expels it via the chimney. The air expelled has to be replaced so a steady flow of cold air into the room results. In really cold climates all stoves should be supplied with air from a source outside the room they are supposed to heat. The air can be drawn from the basement or crawl space below the floor or simply ducted in through the wall by means of flue pipe. Lead the air into the fire from below via the grate.

Cooling a house, if you avoid power-guzzling air conditioners which actually *give off more heat than they get rid of*, depends entirely on house design. The properly insulated house with crawl space below, a shade giving overhang on the roof to the south and a set of screen openings under the eaves to let hot air escape from the interior will remain comfortable even in the hottest weather.

10. Water supply

A house must have water as a body must have blood. The true Whole-earther, determined to simplify life, may carry his water to the house from a spring or stream. But water is heavy stuff and there are easier ways of transporting it. Here are some.

Gravity flow. If you have access to any source of water be it stream, spring, pond or lake, *above* your house, then you have few problems. Gravity will do the work for you and gravity is a force you can rely on. All you need is enough pipe to carry the precious fluid and a reservoir in which it can accumulate. If the water carries sand or silt you will also need a filter bed.

Unless you refuse absolutely to touch the products of modern technology you should use polyethylene pipe for water conduction. It is cheap, durable, easy to handle, can be left on the surface but lasts longer if buried. In cold climates burying the pipe below frost line is essential anyway. PVC (polyvinyl chloride) pipe is better than polyethylene but more expensive. An underground concrete storage tank can be built to hold 10,000 gallons either using Ken Kern's technique of the rotating spiral form or using conventional wooden forms and making the tank rectangular. The filter bed, if necessary, should be placed ahead of the storage tank. Your storage tank should be about 20 feet above the house to give adequate water pressure. If you wish to obtain a water reservoir without building for eternity, try a plastic swimming pool, prepare a proper cover and get the water out by means of a siphon (no good for cold climates).

If you have a spring above the house you should protect it with a spring house. This will provide a reservoir for the water and help to prevent contamination. Springs are generally slow-flowing so by building a reservoir you make possible a steady accumulation of water against the time when it will be needed.

A reservoir can also be used to accumulate rain water which would otherwise run off the roof of the house and be wasted. These cisterns, if they are to last for any time, should be constructed of concrete and sunk below ground level. It is desirable to place a filter bed between the cistern and the roof, also to keep gutters clean of leaves and debris which will give the water a bad taste.

Raising water. If the source of water is *below* the house the Whole-earther can choose between several methods of raising it to the required level, in addition to the old-fashioned hand pump:

(1) Hydraulic ram. This method requires flowing water. It converts part of the energy of falling water into a pumping action. It will give you about 25 feet of lift for each foot of fall and requires at least 3 gallons a minute of flow. These gadgets are made by Rife Hydraulic Engine Manufacturing Co., Box 367, Millburn, N.J. 17041.

(2) Reciprocating wire power. This method has been used by the Amish of Pennsylvania. A small undershot waterwheel (diameter 1 or 2 feet) is located by a stream. The wheel shaft is fitted with a crank which is attached to a triangular frame pivoting on a pole. The reciprocating action is transmitted to a pump by a series of wires kept tense by counterweights.

(3) If you have neither spring, lake nor stream and your rainfall is not sufficient to fill a cistern you may have no alternative but to dig for water. Water that sinks into the soil is carried in special layers called *aquifers*. These are porous rocks, layers of sand or gravel. The top of the water level in these aquifers is called the water table. The level of the water table varies from year to year and at different times of the year. A well, to be effective, must go below the level of the water table at its lowest. The amount of water that can be taken from a well depends on the quality of the water table and the rate at which water flows within it. An aquifer of sand or gravel will yield as much as 3,000 gallons per minute but a limestone aquifer will yield between 10 and 50.

The first question anyone confronts who plans to dig a well is where to dig. Unless he prefers to rely on dowsing (concerning the efficacy of which there are conflicting opinions) he can use three criteria for locating his well: (a) Nearness to surface water. You are likely to strike water near a pond, lake or stream even if the stream dries up completely in summer. (b) Topography. Groundwater gathers in low areas, therefore the lowest place is the best place to dig. (c) Sediment type. Sand and gravel are porous, clay is not. A study of sediment types and their relationships may help you to find the aquifers (fig. 29).

The next problem a well maker confronts is how to get down to the aquifer. There are two methods: (a) Dig down bodily, which involves making a hole wide enough for a man to work in and hauling out the dirt. (b) Drill the hole, either by hand or machine.

That excellent publication *VITA* (*Village Technology Handbook*, College Campus, Schenectady, N.Y. 12308) points out that dug wells are always more expensive to build than drilled tube wells

HOUSING 171

and are limited as to the depth to which they can penetrate. Also a dug well can easily cave in on the digger unless special precautions are taken. Wherever possible, therefore, a tube well should be used rather than a dug well.

You can drill your own tube well if you locate in an area in which a simple earth auger can be operated (such as alluvial plains with few rocks in the soil). The aquifer should be 50 to 80 feet below the surface. The storage capacity of a tube well is

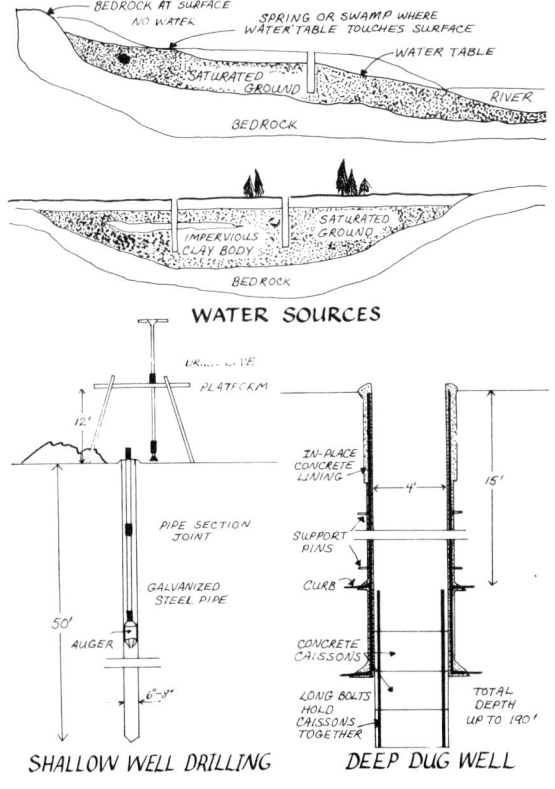

FIGURE 29

small but you can drill several of them with the same expenditure of energy as it takes to make one dug well, which increases your chance of hitting a reliable aquifer.

Here is a simple method of hand drilling a well, used by the American Friends Service Committee in India. Procure or make an earth auger from 6-inch steel tube (fig. 29). Different types of augers suit different kinds of soil. The difference is in the angle of the cutting edge. Attach to your auger a 10-foot length of 1-inch-diameter galvanized steel pipe. Build a platform 10–12 feet from the ground above the point where you wish to drill. Start the hole, taking care to make it vertical, using two short lengths of pipe and a coupling to make a handle. A tripod with rope and pulley above the platform will facilitate lifting out the auger for emptying. As the hole gets deeper, add another length of pipe. Continue drilling until an aquifer is penetrated.

Two problems may be encountered: (1) You hit a rock. In this case you must either abandon the well or knock a hole in the rock. To do this you will need a mild steel bar about 5 feet long and 2¾ inches in diameter, weighing about 175 pounds, tipped with stellite (a very hard type of tool steel) with a handle at the top to which a rope can be attached. This rope is run through the pulley above the well, and by means of it the drill is alternately raised and dropped. To do this one needs five to seven helpers. However, by jacking up the rear wheel of a car and replacing the wheel with a small drum you can make a friction device which will lift the bit for you. (2) You hit quicksand. This is a mix of sand and water that flows and fills in your well. The solution is to lower casing into the well and continue to drill with a narrower auger.

Once you have hit water you will probably need casing anyway. The casing serves two purposes: At the top of the well it serves to prevent groundwater from polluting the well water. The top casing is therefore impermeable. At the bottom of the well it prevents cave-ins. The casing below water level is therefore perforated with small holes. Black plastic pipe can be used for casing for narrow wells. For wider ones asbestos cement units are inserted.

Once you have hit the aquifer your auger will no longer bring the soil to the surface because that soil will be turned to mud of the consistency of pea soup. To continue drilling you will need a "bailing bucket." This can be made from a piece of steel pipe 71 inches long and about ½ inch smaller in diameter than the auger, closed at one end, open at the other. It is dropped into the well until filled with mud, raised and emptied. It is a good idea to construct a flume to conduct the mud away from the drilling site.

If for some reason you cannot make a drilled well and decide to dig, then line it as you go. In this way you avoid the risk of being buried alive or losing much of your well through cave-ins. For one man digging, the hole should measure 3¼ feet, for two 4¼ feet. The top 15 feet of well can be held with a cast-in-place lining. The remainder can be protected by 3-foot-diameter caissons bolted together, which fall as the soil is dug away beneath them (fig. 29). The precast units are, of course, very heavy and a block and pulley is needed to facilitate handling. But once such a well is dug it should last forever.

11. Waste disposal

No matter how frugally he lives the Whole-earther's body will produce waste in the form of urine and feces. They must be disposed of somehow and the choice is wide, ranging from shitting over a hole in the ground and burying the feces to installing a water-guzzling (4 gallons a flush) toilet with a septic tank and drainage field.

In choosing his method of waste disposal the Whole-earther will be guided by two considerations: (1) Fecal matter should not be left exposed or be disposed of in such a way that it can contaminate drinking water. A number of disease entities, ranging from viruses to parasitic worms, can inhabit human feces, hence the need for care. (2) Both feces and urine contain the nutrients which man derived from plants and if man is to complete the cycle of nature these nutrients should be returned to the garden. The ideal waste disposal system therefore has two attributes: it will render the waste safe by destroying the pathogenic organisms; it will restore the wastes to the soil in a form which plants can use.

The house builder, in particular, must make up his mind right from the start whether his waste disposal facilities will be in the house or outside it. Because it influences building design, the inside waste disposal unit will be considered first.

The kind of waste disposal unit imposed on the house owner by bureaucrats through building codes has three parts: a flush toilet, a septic tank and a drainage field. It can be criticized on three grounds: it is expensive, wasteful and, on heavy clay soil, impractical. The drainage field won't drain.

In designing a better waste disposal one must consider the following points:

(1) The ordinary flush toilet is designed without any consideration for the realities of human anatomy. Humans naturally defecate in a squatting position. This exerts the maximum pressure on the contents of the abdomen. There is no need to support the buttocks on a pierced seat unless one is in the habit of using the toilet as a place for meditation, reading the newspaper or escaping from the family. But defecation and meditation are two quite different functions. Defecation is an athletic exercise and a very important one as the regular elimination of waste is vital to health. A set-up therefore should be used that encourages the body to adopt the proper posture for this vital function.

Ken Kern, with his usual practicality, has taken this fact into consideration in designing the all-purpose washing and waste disposal unit (fig. 30). The toilet bowl in this unit is cast in a mold which can be designed either to flush forward or backward. The dimensions of the mold are as shown. Bowl and trap must be cast separately. Ken Kern has obtained the master molds from the Thailand Ministry of Health and is willing to loan them to people interested in building their own toilet bowls.

(2) The usual method of waste disposal (a septic tank) does not enable the Whole-earther to return his wastes to the garden or to get the benefit of the gas they generate (chiefly methane), which can be used for cooking or heating.

HOUSING 175

In the cycle of nature there are two processes which transform wastes and return them to circulation. The first is aerobic (needs oxygen), the second is anaerobic (takes place only in the absence of oxygen). The compost heap is aerobic and the septic tank is anaerobic.

The most simple and safe way of returning human waste to the soil is to compost it along with vegetable and other animal wastes. A well-made compost heap attains a temperature of 160° F, hot enough to destroy any disease organisms. The com-

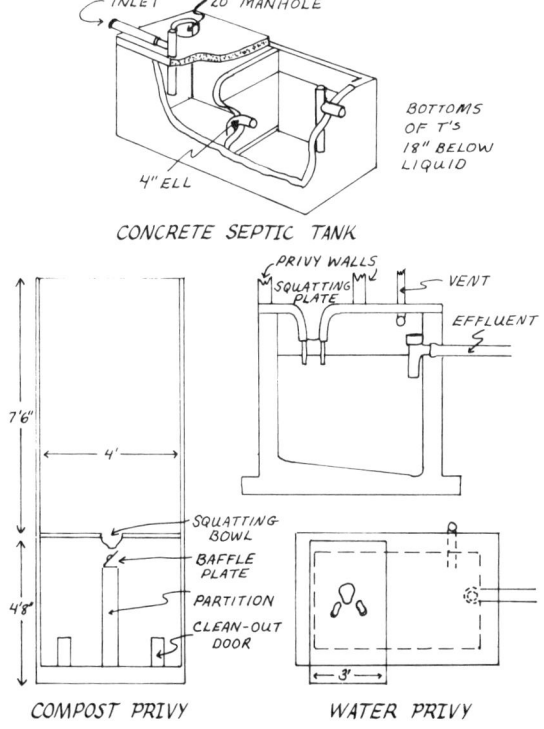

WASTE DISPOSAL UNITS

FIGURE 30

posting of human waste is possible only if they are kept reasonably dry. It is not practical if one uses a flush toilet that dilutes them with 4 gallons of water every time it is used. An old-fashioned privy using a container partly filled with earth, in which the waste is covered with a mixture of earth and slaked lime, is reasonably safe, provided enough earth is used to cover the shit properly and prevent the inroads of flies, which are the worst carriers of disease. Such a privy is easy to construct but it needs emptying with annoying frequency and careless habits can leave it open to fly invasion. A far better device is the two-chambered compost privy combined with the Thailand-type flush bowl (fig. 30). This is recommended by Ken Kern and by the Indian Council of Agricultural Research at Bangalore. The following points should be noted:

(1) The whole unit with its two storage chambers should be either cast in concrete or built from cement blocks and properly waterproofed. The water seal in the squatting bowl will prevent both ingress of flies and egress of odor.

(2) The size of the storage compartment depends on the number of people using it. The compost unit can be partly filled with sawdust or straw to help absorb the urine. Kern calculates that a family of five will fill a compartment 1 cubic yard in volume in about 9 months.

(3) After one side of the privy has been used for 9 months the baffle should be switched over and the other side used. The excreta in the first compartment should be left for about 6 months to ripen. Addition of powdered soybean or an actively growing culture of yeast will hasten the process.

(4) After 6 months the ripened compost can be removed. It should be fully digested and more or less odorless. It cannot, however, be guaranteed free of disease organisms. To complete digestion and sterilize this material it should be used along with other animal waste in the building of a proper compost heap, the heat of which will kill the disease germs. It can then be safely spread on the garden and the Whole-earther will have the satisfaction of knowing that the cycle has been completed.

The anaerobic disposal unit works on a different principle. It dilutes the waste with a large volume of water, holds it in a waterproof, airtight container until it breaks down. The overflow runs into a drainage field and returns to the soil. Part of the waste breaks down into sludge which accumulates and must be removed at intervals. A simple type of water privy in which the squatting plate is immediately over the container and connected to it by a drop pipe is shown in fig. 30. This unit should be connected to a drainage pit. The amount of effluent can be estimated at about 2 gallons per person per day. A more elaborate system in which the toilet is remote from the septic tank is shown in fig. 30. The building code demands are shown in table III.

TABLE III
Building code requirements for septic tanks

Number of bedrooms	Inside dimensions, feet			Thickness of concrete, inches				Capacity, gallons
	Width	Length	Depth	Sides	Ends	Floor	Top	
1–2	3½	8	5	5	5	5	8	800
3	4	9	5	5	5	5	8	1,000
4	4	10	5	5	5	5	8	1,200

12. Houses on water

As the land grows more crowded, men may have no alternative but to move onto the surface of the ocean, which covers about three-fourths of the globe. Such a move has occurred already in densely populated areas like Hong Kong, where a large proportion of the population live in the harbor on the water. The crowding and the level of pollution are horrendous. This does not bother the water dwellers, who have a well-developed immunity to the various disease organisms they pour into the harbor. But it is perfectly possible, by using a little intelligence, to live on the water without polluting it, and this, for some, may represent the ideal way of life.

There are two ways of living on the water: in individual houseboats, and in floating communities which I propose to call *aqua-*

coms. The aquacoms are in the future. The houseboats are already with us and will be described first.

Houseboats. Almost any boat can be a houseboat. It is perfectly possible, as people like the Monforts have shown (*The Mother Earth News*, no. 16, p. 69), to live on a 30-foot fiber glass sloop and be reasonably comfortable. Indeed, if you don't mind being a bit confined, a deep-water vessel able to go anywhere is a better buy than an orthodox houseboat, which can cost anywhere up to $24,000 and is seaworthy only in sheltered coves.

The would-be boat dweller has two alternatives: buy a new or used boat, or build a boat. Buying a new boat is bound to be expensive. Buying a used boat can be very profitable if one knows enough about boats to distinguish between defects that can be corrected and defects that are past repair, short of rebuilding the whole hull. As the hull is all you will have between you and a watery grave it stands to reason you will want a good one. In a wooden boat there are three parts to a hull: the stem, ribs and planking. Replacing ribs is difficult, replacing sections of stem (which is the boat's backbone) is even more so. Planking can be replaced more easily. It is worth going over an old boat inch by inch with an ice pick, probing for soft spots.

If you aren't particular about appearances an old fish boat or tug may be better than a sailboat. Fish boats of the kind that go out on the open ocean are designed to take rough seas. They are broad in the beam and offer plenty of room. Tugs are also good potential houseboats, being well constructed, roomy and sometimes already equipped with living accommodations. For useful tips, see "How We Found a Live-Aboard Boat in B.C." by C. Houff, G. DuLac and J. Weierman, *The Mother Earth News*, no. 16.

It is, of course, quite possible to find a used houseboat, especially in the British Columbia area around Vancouver Island. The problem here is checking the condition of the pontoons on which the vessel floats. Unless you have scuba gear this is hard to do. Of the various materials from which pontoons can be built, marine aluminum is probably most durable, closely followed by ferrocement, fiber glass and steel. Remember a houseboat, being commonly quite unseaworthy, will limit your movements even if it has a motor. They are good for lakes or protected inlets. (For further material, write *Family Houseboating,* P.O. Box 2081, Toluca Lake, Calif. 91602.)

Building one's own live-aboard boat is certainly more instructive than buying it. The process will give the true Whole-earther insights into an ancient and honorable art that goes back beyond the time of Noah. The amateur boat builder will probably be attracted by two modern materials, ferrocement and fiber glass. They are easier to work with than wood and they have very obvious advantages. Ferrocement needs no painting, is not attacked by ship worms and gets stronger as it cures. Fiber glass is light and durable and can be easily applied over a plywood hull, to which it will bond readily.

Both these materials are rather tricky to work with and the amateur boat builder is advised to get some tips from an expert before starting anything larger than a rowboat. Ferrocement boats are constructed on a wooden mold made of flexible laths nailed onto station frames, each of which represents a cross section of the boat. The cedar strips are then covered with a layer of polyethylene. The mold is covered with 20-gauge, 1-inch-mesh chicken wire (four thicknesses) over which the rebars (reinforcing rods) are stapled to the frame. Horizontal rods are spaced 2 inches apart, vertical rods are spaced on 6-inch centers. Over the rods two more layers of chicken wire are applied.

This part of the construction can be done in a leisurely way on weekends or after work. But the actual plastering requires effort of a higher level of intensity because it must all be done in a single operation, and there must be no mistakes. For this reason, though many helpers can be employed to mix the cement, a professional plasterer should be hired to do the actual application. Cement is unforgiving stuff and mistakes are hard to rectify. (For plans, advice and sites on which to build, consult Samson Marine Design Enterprises Ltd., 833 River Rd., Vancouver, B.C., Canada.)

Aquacoms. The aquacom is an integrated floating community. It can be built with the technology already at our disposal. By the use either of the floating pylon or the floating box a foundation can be laid sufficiently stable to withstand the motion of the ocean. The pylon technique, which uses concrete or metal tubes up to 250 feet in length floating deep in the ocean, is not practical for the small-scale builder. The concrete box is more attractive, being essentially a box-shaped ferrocement hull which can be linked to other such hulls to form a raft on which buildings can be constructed.

A number of problems have to be solved before such aquacoms become practical realities. To solve them it will be necessary to create small experimental units in environments such as the water off the Hawaiian Islands. The population explosion in Hawaii and the soaring cost of real estate makes the aquacom a practical proposition for this state. What we need is some enterprising group willing to experiment with this ecosystem and discover to what extent it can be self-contained and non-polluting. Elaborate plans have already been worked out for floating cities off the coast of Oahu but the ecological impact of such aggregates (especially the problem of recycling waste) seems to have been overlooked.

Four questions must be studied before the self-supporting aquacom can become a reality: (1) Can the aquacom draw all its food from the ocean? (2) Can the aquacom recycle its wastes in such a way as to avoid polluting the environment? (3) Can the aquacom, by using solar distillation and trapping rain water, obtain enough fresh water for all its needs? (4) What is the cheapest and most stable design for a small aquacom of about 20 individuals?

We have every reason to believe that the answer to question 1 is yes. Animal food could be provided with suspended cultures of oysters or mussels and fish farming or fish hunting. Perhaps a herd of domesticated female porpoises or small whales could be trained to provide milk. The milk of these aquatic mammals is astonishingly rich though milking them might be a problem. Problem 2, the recycling of wastes, is fairly simple to solve. The digested sewage can be used in part to generate methane for cooking, the residues can provide a culture medium for unicellular algae which in turn can be food for oysters. Problem 3 can be solved even in areas of low rainfall by the installation of a sufficient area of solar stills. A solar still of the plastic film type produces about 1 gallon of pure water per 12 square feet per day. This can be partly diluted with sea water, which restores some of the minerals lost in distillation. Collection of rain water in a floating cistern will supplement the supply.

Problem 4, which concerns the cheapest and most stable design for small aquacoms, is the one most urgently in need of solution. Imaginative architects like Kiyonori Kikutake, Paoli Soleri and Buckminster Fuller have all offered designs for floating cities. "Cities of tomorrow will rise from the sea and float on its surface like water lilies. They will grow, ripen and flower for the sole purpose of lifting the spirit of man. And when they have outlived this reason for their being they will vanish into the watery depths that bore them" (John Lear, "Cities on the Sea?," *Saturday Review*, Dec. 4, 1971, p. 80). Students at the University of Hawaii, guided by a naval engineer, John Craven, took a step in this direction, designing a module consisting of three hollow pylons 450 feet long, 200 feet of which would hang beneath the water. But the cost of this floating foundation, bare of buildings, was estimated at $10 million. Twenty such modules, enough for a small floating city, would cost $200 million. Expensive even by Hawaiian standards.

More practical for the small aquacom is the box of prestressed concrete designed by Ben C. Gerwick, Jr., of San Francisco (see his *Construction of Prestressed Concrete Structures*, Wiley-Interscience, New York, 1971). The floating box is a proven device, employed in the invasion of Normandy in World War II. A group of aquatic Whole-earthers could certainly use this technique for building a modest aquacom. It would be an interesting experiment to start such a venture.

III HEALTH

In the temple of the great Aesculapion of Cos these words of Heirophilus of Alexandria were inscribed in marble: *Without health wisdom is darkened, art eclipsed, strength disabled, riches worthless and reason impotent.*

The saying is as true today as it ever was. The ancient Greeks, though they had many erroneous ideas about the forces that induced disease, had some good ones about the nature of health. Health depended on the balance between various forces working in the body. When these forces were harmonized there resulted a state of well-being (*eucrasia*), when they became disharmonized there resulted dis-ease (*dyscrasia*). Man needed to be in harmony with himself and in harmony with his surroundings. Such harmony was the basis of the sacred *ataraxia*, which was the root of good health both of the mind and of the body.

To enjoy the full measure of good health the Whole-earther must take steps to regain the state of harmony. To do this he must first train his awareness. Primitive man, the hunter and forager, had to have an intense awareness of his surroundings. If he wandered around in a fog of dreams he starved. His prowess as a hunter and a warrior depended on his watchfulness, his ability to pick up small clues from his environment. He was constantly faced with challenges and his life depended on his response to these challenges.

The true Whole-earther who has fled from the vast air-conditioned termitary and imposed on himself a more exacting way of life must also, if he wishes to survive, cultivate awareness. Through this awareness he can learn how to keep his various bodily functions harmonized. He can train and toughen his body through the strenuous activities that a down-to-earth life style provides. His way of life will demand that he fell trees, spade soil, build, harvest, hunt, fish. From all these activities he can learn how to breathe deeply, how to use his muscles, how to move, how to remain motionless, how to be alert and how to relax. He will learn the old skills that "primitive" man knew and that the contemporary city fat cats have lost.

But although this life style will generally ensure good health both for the Whole-earther and his family it is necessary to be prepared for illness and accidents. Much ill health can be prevented if appropriate steps are taken in advance. It is desirable that the Whole-earther be able to recognize the commoner forms of illness and also that he know what steps can be taken to prevent illness from developing.

HOW THE BODY WORKS

Anyone who wishes to live independently needs to know something about how the body works. It is ludicrous in this age of science to live in ignorance about the working of one's own organism. Every true Whole-earther should have enough medical knowledge to be able to distinguish dangerous illnesses, which need medical care, from less serious conditions, which can be cared for at home. He also needs to know what to do in an emergency.

A healthy body can take care of itself but every body has certain strengths and weaknesses which impose limits on its workings. It is necessary to know one's physical limitations. Some limitations can be overcome by training; others cannot. It is necessary to distinguish between the two.

The human body is a chemical factory. It takes in certain raw materials and transforms them, using the energy derived from these transformations for movement, thought, love, hate and all the complex manifestations of life. All these chemical transformations take place in certain systems of organs. To know how the body works one must know something about how these systems are made and what they do. One must also know how they can go wrong and what can be done to put them right.

1. Digestive system

It is logical to consider the digestive system first because it was formed first. All animal organisms above the level of the jellyfish are essentially tubes. They take in food at one end of the tube, process it, absorb a portion of it and reject the waste material at the other end of the tube (fig. 31). If anything goes seriously wrong with the tube the organism will perish. In man the tube begins with the mouth and it is in the mouth that the process of digestion starts. The first organs of digestion are the teeth, which grind the food. They are wretchedly vulnerable entities and certainly represent one of our worst evolutionary blunders. The birds, which use stones in a muscular crop to grind their food, are obviously better off. So are the sharks, whose teeth can be replaced. But we are stuck with one set of grinders which must last from early childhood to old age, which are prone to rot and are equipped, quite needlessly, with sensory nerves that give us agonies of toothache. Decidedly, if man wanted to redesign his body, he might start with his teeth.

Teeth grind the food and mix it with the first of the digestive juices, the saliva. All digestive juices have one thing in common. They contain enzymes which break complex substances into simpler ones. The saliva acts on starches and turns them into sugar. Needless to say, it can't do much acting if food is bolted down in a hurry. Eating can be done well or badly, and is generally done

HOW THE BODY WORKS   185

badly, without attention or awareness, without gratitude to the cosmic process that generated the food, and without enough chewing to reduce food particles and mix them thoroughly with saliva. The habit of talking during meals is especially deplorable. Eating is an important function and cannot be done properly if one insists on jabbering at the same time.

After the food has been ground and mixed with saliva it is forced into the esophagus by a special set of muscles. Once it is swallowed it passes out of conscious control. From then on, the

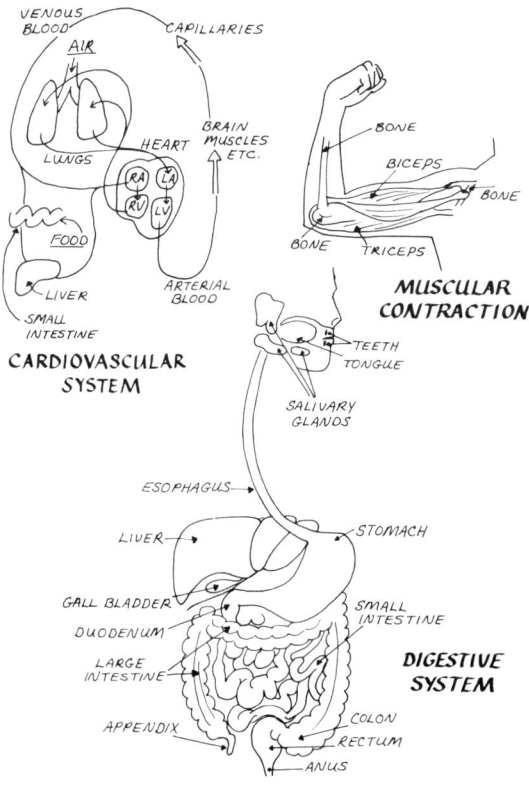

FIGURE 31

processing of the food is managed entirely by the instinctive brain.

The esophagus passes the food along to the stomach by a series of contractions. In the stomach the food is mixed with hydrochloric acid strong enough to curdle milk. Food is held in the stomach and released in small amounts into the duodenum, where several enzymes act upon it. Fats are acted on by bile, which comes from the liver and is stored in the gall bladder. Proteins are broken into amino acids by enzymes produced in the pancreas. Starch is broken down into sugar. All this breaking down is necessary because no food can be used in the body until it has been reduced to soluble form.

The food is taken up by the blood as it flows through millions of finger-like projections (*villi*) that line the walls of the small intestine. This 20-foot-long twisted tube is the vital part of the digestive system. The large intestine is not much more than a storage organ that concentrates waste material by removing water and compressing the material into feces, passing it into the rectum from which it is expelled via the anus.

2. Cardiovascular and respiratory systems

These systems have several functions. The heart keeps the blood flowing through the body by its rhythmic contractions. The lungs supply the blood with oxygen and remove from it carbon dioxide. The arteries transport the oxygenated blood to the tissues. The thinner-walled veins transport the blood back to the heart for re-oxygenation in the lungs (fig. 31).

In addition, the blood picks up food from the small intestine and passes it to the liver, which is the body's chief chemist. From the liver the blood flows on to carry food materials to every cell in the body. So the blood must transport two substances, oxygen which it gets from the lungs and food materials which it gets from the small intestine and the liver. The blood, on its return journey, must carry waste material to be removed by the kidneys and carbon dioxide to be removed by the lungs. The blood also carries a portion of the body's defenses against invasion in the form of white blood cells and various antibodies as well as the hormones produced by various glands. Blood is indeed a complex and versatile substance.

The heart, which pumps the blood, is a double organ. The right heart receives blood from the veins into its first chamber (right atrium). The first chamber contracts, forces the blood through a valve into the second chamber, the right ventricle. Then the right ventricle contracts and sends the venous blood to the lungs, in which it is cleansed of carbon dioxide and recharged with oxygen. The right heart is entirely occupied with circulating blood through the lungs.

The blood returns from the lungs to the first chamber of the left heart (left atrium). This chamber contracts and forces the blood through a valve into the largest, strongest chamber of the heart, the left ventricle. This chamber then contracts and forces the blood into the arteries, from which it returns to the veins via small blood vessels called capillaries. These minute tubes are so numerous that their total length would be 60,000 miles, or 2½ times the distance around earth's equator!

The lungs, which aerate the blood, are made up of tubes. The first and biggest tube is the trachea, or windpipe, which divides and sends one branch to each lung. The tube divides into finer tubes (*bronchi*) and still finer tubes (*bronchioles*) ending in air sacs (*alveoli*). In the walls of these sacs the blood circulates, gives off part of its carbon dioxide and takes up some oxygen. It uses for this purpose the red coloring matter carried by the red blood cells (*hemoglobin*). Hemoglobin is bright red when charged with oxygen, bluish when deprived of it. Hence the difference in color between arterial and venous blood.

There are really four circulations going on in the body. The flow from heart to lungs and back to the heart is called the *pulmonary circulation* and its purpose is to aerate the blood. The flow from heart to muscles, brain, etc., and back to the heart is called the *systemic circulation*. It supplies the organs of the body with food and oxygen. The flow from the small intestine to the liver is called the *portal circulation*. It conducts food material to the liver, where it is processed and stored. The flow of blood through the kidneys is called the *renal circulation*. It filters toxic wastes out of the blood and concentrates them in the urine.

3. Nervous system

The nervous system links the body to its surroundings. By means of the *sensory* nerves it tells us what is happening out there. By means of its *motor* nerves it tells us what action to take. The sensory and motor nerves alone would not be much use to us without the interpretive organ, located in the brain. This interpretive organ receives the information that pours into the sensory nerves. It compares the information with other information stored in the memory. On the basis of this comparison, made with lightning rapidity, it issues orders to the motor nerves. If a dog approaches us wagging its tail, the interpretive organ says, "Friendly doggy. Pat it on the head." But if the dog approaches snarling and barking it says, "Savage beast. Attack it or run away."

This part of the nervous system which works more or less through the conscious mind is called the *central* nervous system. But there is another part of the nervous system not under our conscious control. This is called the *autonomic* nervous system. For example, when the snarling dog is interpreted as a threat and orders are issued to attack it or run away a series of changes takes place to help the body cope with the crisis. Blood is withdrawn from the skin (we turn pale), adrenalin, the hormone of action, is poured into the blood, the heart rate and breathing rate both increase. The body is thus put into condition either to fight or flee without the conscious mind being directly involved.

The nervous system is made of nerve cells. These cells are the most highly specialized in the body and, if destroyed, are not replaced. We have only a limited supply and they have to last a lifetime. The long nerve fibers that go out from these cells are combined into bundles which are white in color and spread to every organ in the body. These white threads are the nerves. Our muscles are useless without the nerves to stimulate them to action and a muscle deprived of its nerve supply soon withers. From the brain located in the skull and wrapped in the *meninges* the nerves go out to the various parts of the body. The grand highway of the nervous system is a hollow in the center of the backbone into which the nerves are packed in the form of a cord about 1½ feet long and slightly thicker than a pencil (spinal cord). The cord nerves which carry messages *from* the brain to muscles (*efferent* nerves) emerge from the front part of the cord. Those that carry messages *to* the brain (*afferent* nerves) enter the back part of the cord. The nerves have to enter the cord through gaps between the vertebrae that make up the backbone.

Man and other animals go through life headfirst. This means that all the most important sense organs, those of sight, hearing, smell and taste are located in the head piece. The nerves that carry this information go direct to the brain. These are the cranial nerves and the amount of information they carry is prodigious. Only a fraction of it reaches the conscious mind, which is protected by filters that remove irrelevant information. Without these filters we would be overwhelmed by the sheer volume of our sensations. During the hours of sleep these sensations are filtered out entirely or reach us obscurely in the form of dreams.

The autonomic part of our nervous system, over which we have no direct control, is constantly affected by our conscious life. It sends its branches to all the main organs of the body and either stimulates them to action or slows them down. Its effect is especially noticeable on the circulatory and digestive systems. All sorts of novelists' cliches—"He turned red as a beet, white as a sheet, his heart pounded like a sledge hammer, he had butterflies in his stomach, etc."—refer to the action of this part of the nervous system. In modern man it tends to be overreactive and many ills of civilization, ranging from high blood pressure to gastric ulcers, are due to its overactivity.

4. Muscular system

The muscle is an engine that converts energy stored in certain chemical substances into movement. Muscles act only by contracting. They can pull but they never push. The body contains three kinds of muscle. The *skeletal,* or *voluntary,* muscles are under our conscious control and make possible our voluntary movements. The *involuntary* muscles are found in the walls of the intestine, blood vessels and glands. They differ in appearance from the voluntary muscles and are not under our control. The *heart* muscles are in a class by themselves. Their rhythmic contractions are built into the muscle cells. They go on contracting even in tissue culture after being completely separated from the animal. Their rhythm of contraction varies from 1,000 times a minute in a canary to 25 times a minute in an elephant. Their contraction rate is regulated by the autonomic nerves.

The voluntary muscles work in pairs, one undoes what the other has done. To produce movement the muscles must have something solid to pull against and this solid anchorage is provided by the bones. Nearly all the voluntary muscles are strung between two bones and act by pulling one bone in relation to the other. A simple example is the flexing of the arm. The biceps is attached to the shoulder blade at one end and the bone of the forearm at the other (fig. 31). When it contracts, the forearm is pulled closer to the shoulder. To reverse this movement the triceps on the lower side of the arm must contract to pull the forearm down again. Efficient muscular activity therefore always involves contraction of one muscle or set of muscles and the simultaneous relaxation of their opponents. Very tense people are unable to relax properly when moving and so waste a lot of energy in forcing their muscles to work against each other.

All muscles tend to shrivel when left unused and conversely all muscles tend to develop when used frequently and vigorously. Every muscle has its own blood supply and its own nerves. Without the nerves it cannot contract, without the blood supply it rapidly becomes poisoned by its own waste products (chiefly lactic acid). This produces muscle cramp and muscle stiffness. Muscles are sensitive to heat, which tends to relax them, hence the importance of "warm-up" to the dancer and athlete.

The attachment of muscles to bones is by means of *tendons* or *ligaments*, which are the "ropes" that hold the body together. These ropes are very important. They not only transmit the action of a muscle, they also, being elastic, protect the muscle against sudden strain.

5. Skeletal system

Man's skeleton is the frame on which his body is constructed. It is made of bone, a remarkable living tissue of extraordinary strength and lightness (the femur, or hip bone, will withstand compression from 1,800 to 2,500 pounds). Bone is built by special cells called *osteoblasts*, which lay down spicules of mineral material, mainly calcium phosphate. The spicules are laid down in layers, the grain of each layer being deposited at a different angle from that of the next. This results in the formation of a tough laminated structure somewhat like plywood, the spicules being held together by a fibrous cement.

There are more than 200 bones in the human body and they vary enormously in shape, size and function. They fall into six groups:

(1) The skull. The bones of the skull are flattish, providing a box of bone to hold the brain and support the facial muscles. Most of the other bones in the body can move in relation to one another. The bones of the skull cannot. Only the bones of the lower jaw are hinged.

(2) The spinal column. This is made up of 33 vertebrae stacked in a column and held together by powerful muscles. The column can be divided into four parts. The *cervical*, or neck, region carries the skull, the *thoracic* region carries the ribs, the *lumbar* region gives support to the viscera, and the *sacral* region, with its fused bones, holds the hip bones together at the back.

(3) The shoulder girdle and arms. The arms are anchored to the body by two bones, the scapula (shoulder blade) and clavicle (collarbone). To these are attached the long bones of the arm (humerus, radius and ulna), at the end of which are the bones of the wrists and hand.

(4) The rib cage. This cage of curved slender bones encloses the lungs and heart.

(5) Pelvic girdle. This is a massive ring of bones that supports

the viscera and, in the female, holds up the fetus until it is ready to be expelled.

(6) The leg bones are attached to the pelvic girdle. These include the longest, strongest bones in the body (the hip bones) to which are attached the lower leg bones, terminating in the complex bones of the ankle and the numerous small bones of the feet and toes.

Bones that move in relation to each other have a special fluid (*synovial fluid*) that acts as a lubricant. The joint itself is lined with smooth cartilage that has a beautiful luster, polished like a pearl. Bones at a joint are held together partly by muscles, partly by ligaments. In addition to supporting the body, bones house the *bone marrow*, which is concerned in the manufacture of the cells of the blood.

6. Endocrine system

This system is made up of glands that supply various hormones to the body. The master gland is the *pituitary*, at the base of the brain, directly linked to the old brain and under its control. The pituitary acts like the conductor of an orchestra. It tells the other glands when to start operating and how much they should contribute of their special hormones. The pituitary does this by means of hormones of its own which act on certain glands and are labeled *tropins* for this reason. So the hormone acting on the gonads is *gonadotropin*, the one acting on the cortex of the adrenals is *adrenocorticotropin* (ACTH), the one regulating the growth of the body is *somatotropin*, the one acting on the thyroid is *thyrotropin*.

The role of the pituitary is clearly seen when a child reaches puberty. The quiescent sex glands, testes in the male, ovaries in the female, are roused to action by the pituitary hormones and begin secreting hormones of their own. The *testes* produce the male hormone (*testosterone*), the *ovaries* produce the female hormone (*estrone*). These hormones bring about the changes in the body associated with puberty.

The pituitary also produces a hormone that regulates the water balance in the body.

The *thyroid* gland is located in the throat just below the voice box. The hormone it makes (*thyroxin*) controls the rate of metabolism of the body, as a thermostat controls the rate of burning of a furnace. Attached to the thyroid are the *parathyroid* glands, which control the balance of calcium and phosphorus in the bloodstream. Situated just above the kidneys are the *adrenal* glands, having two parts, a core and an outer shell. The core makes *adrenalin,* the "fight or flight" hormone that raises blood sugar, increases heart rate, raises blood pressure. The shell (*adrenal cortex*) manufactures a complicated array of hormones called *corticosteroids,* which play various roles in maintaining the vital balance.

The *pancreas,* located at the level of the pit of the stomach, in addition to producing digestive juices, generates the hormone *insulin*. This is made in special cells called islet cells and regulates the uptake of sugar by the cells of the body.

7. Urinary system

This system is composed of the kidneys and two tubes (*ureters*) which empty into the urinary bladder. A third tube (*urethra*) carries the urine out of the body. The kidneys are highly selective filters. Blood passes through them and is cleansed of various impurities, which are then concentrated by the removal of some of the water and passed into the bladder. Urine is stored in the bladder, which is normally emptied when it contains half a pint.

8. Reproductive system

This system in the male consists of two glands, the *testes,* which are normally suspended outside of the body cavity in a sack called the scrotum (fig. 32). The testes are composed of a twisted mass of tubes, the combined length of which has been estimated as a quarter of a mile. In the tubes are formed the sperm cells. Between the tubes are the interstitial cells that secrete the male hormone. The sperm cells are conducted by the *vas deferens* to the *seminal vesicle,* in which they are stored. Two glands, the *prostate* and paired *Cowper's glands,* mix their products with the sperm cells when they are discharged. The total

194 HEALTH

mixture is the semen, which is forcibly discharged by special muscles. One discharge of semen contains about 300 million spermatozoa. Discharge is via the *penis,* an organ containing a spongy mass of tissue that can be distended with blood and thereby made erect.

The female sex organs (fig. 32) consist of the *vagina, uterus, uterine tubes* and *ovaries*. The vagina is covered on the outside by the *labia,* which also protect the *clitoris,* the equivalent of the penis in the male. Unlike the male sex glands which manufacture semen more or less continuously, the female ovaries work rhyth-

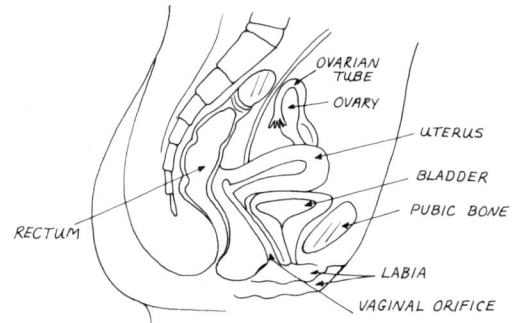

FEMALE ORGANS

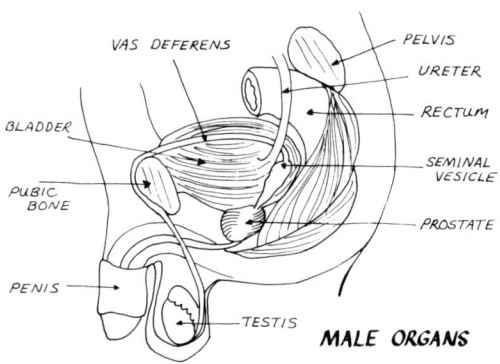

MALE ORGANS

FIGURE 32

mically to produce one sex cell (*ovum*) about every 28 days. Egg production is first from one ovary, then from the other. The egg cell develops in a liquid-filled cavity called a *follicle*. The growth of this follicle is controlled by a hormone produced by the front part of the pituitary (*follicle-stimulating hormone*, or FSH). The follicle gets large and it stimulates the ovary to produce estrogen. This stimulates the tissue lining the walls of the uterus (*endometrium*), which thickens in preparation for receiving the fertilized egg (fig. 33).

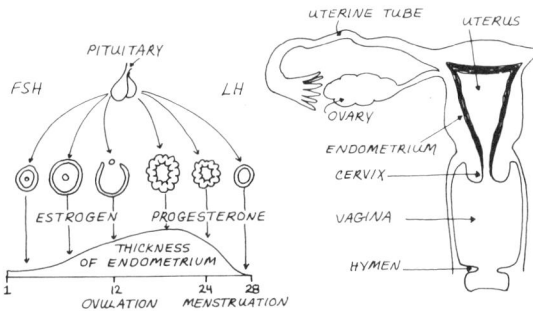

MENSTRUAL CYCLE

FIGURE 33

The follicle finally bursts (generally between the tenth and fourteenth day of the cycle). In so doing, it liberates the egg cell and ovulation occurs, accompanied by a slight rise in the woman's body temperature. The egg cell passes into the uterine tube. Meanwhile the pituitary stops producing FSH and starts to generate a second hormone (*luteinizing hormone,* or LH). The hormone acts on the burst follicle on the wall of the ovary, converting it into a yellow body (*corpus luteum*). The yellow body secretes the second female sex hormone, *progesterone,* which completes preparation of the uterus for the implantation of the fertilized egg.

The egg is fertilized in the uterine tube if it encounters a large enough number of viable sperm. In this event, the egg divides and begins to grow while in the tube, then burrows into the wall of the uterus and establishes itself like a parasite feeding on the mother's blood. The pregnant uterus sends a signal to the pituitary, telling it to maintain production of LH. The corpus luteum grows larger and production of progesterone continues. The progesterone prevents the uterus from expelling its contents until the proper signal is given nine months later. In the later part of pregnancy the pituitary generates another hormone, *prolactin*, which stimulates the development of milk glands in the breasts.

If the egg is not fertilized, the corpus luteum on the surface of the ovary will begin to degenerate, the endometrium lining the walls of the uterus will break down and be discharged via the vagina. This process commonly occupies the last 5 days of the menstrual cycle.

WHAT CAN GO WRONG

The harmonious operation of the body can be upset by various things: (1) improper food intake, (2) intake of poisons, (3) accidental injuries, (4) attacks by parasites, (5) hormone deficiencies or imbalances, (6) breakdown of growth control, (7) breakdown of waste disposal system, (8) breakdown of circulatory system, (9) breakdown of nervous system. Some of these disharmonizing influences can be easily avoided. Others are harder to escape. With age, of course, the body deteriorates inevitably. It is a machine with built-in obsolescence and it lasts, if well cared for, from 80 to 100 years.

The art of living well, as far as physical health is concerned, is to prevent disharmonizing influences from gaining access to the body. This involves the practice of preventative medicine, which is the best form of medicine. If the disharmonizing influences do gain access to the body the next step is to recognize them and do what can be done to limit the damage.

1. Food intake

Malnutrition is a worldwide problem and it takes two forms: malnutrition due to excess, and malnutrition due to deficiency. The harmony of the body can be disturbed by overeating as well as by undereating. Studies with rats have shown that a sparse diet, provided it is properly balanced, produces a healthier though hungrier rat than does a liberal diet. And the rats on the sparse diet regularly outlived their well-fed companions. The thin rats buried the fat rats.

There is, however, an important difference between a sparse diet and an unbalanced diet. A sparse diet provides adequate amounts of all the food components the body needs. An unbalanced diet may provide some components in excess and be deficient in others.

Human food must contain four components: proteins, fuel foods, mineral elements, vitamins.

Proteins. These are big molecules made up of amino acids. It is from these amino acids that the man-body builds itself and with these it repairs itself. Man's body needs 20 amino acids. Eleven of these it can make from other materials. Nine it cannot make and must get from food. Any diet, to be adequate, must provide all nine amino acids in sufficient amounts for body building.

Fuel foods. Fuel foods provide calories but are not useful for body building. Such foods are fats, sugars and starches. They are present in plants and are available in abundance. Potatoes, bananas, yams, onions, root vegetables all contain large amounts of starch. Ripe fruits provide various forms of sugar, peanuts and sunflower seeds offer oil. It is cheap to live on these vegetable foods and many people in poor countries, especially in the tropics, eat them to excess. The result is malnutrition due to protein deficiency, which affects mainly the children and is called *kwashiorkor*.

Mineral elements. The body must have certain mineral substances. The one most commonly lacking in diets is calcium, used in bones and in teeth. Calcium deficiency can be particularly damaging to pregnant women because the fetus draws heavily on the mother's calcium supply. Milk is a good source of calcium. Calcium is also present in most leafy vegetables. Iodine is another mineral element sometimes lacking in food. The lack can be compensated by use of iodized salt or plenty of seafood. Phosphorus is absolutely vital to the body. It not only plays a part in bone formation in the form of calcium phosphate but also provides the vital energy-storing compound of the body (adenosine triphosphate, or ATP). Also, the entire genetic code is written in chemical characters on a backbone of sugar and phosphoric acid. Without phosphorus life as we know it would be impossible.

Vitamins. Vitamins are needed in very small amounts. They form part of the make-up of enzymes, the vital catalysts that control the body's chemistry. The man-body cannot make these chemicals so must get them from outside. One group of vitamins is soluble in water, the other group is soluble in oil. The water-soluble group includes vitamin C (ascorbic acid), a rather simple substance present in orange juice and in most fruits and vegetables. Some animal bodies (the rat body, for instance) can make their own vitamin C, but the man-body cannot, and cannot even store very much. So a lack of vitamin C quickly leads to body sickness that manifests as scurvy, that scourge of the ancient mariners.

The other water-soluble vitamins, ranging from B_1 to B_{12}, are all present in whole grains, leafy vegetables, fruit. One of the diseases associated with a lack of these vitamins (*beriberi*) is a direct result of man's tendency to overrefine his food. Beriberi was first observed in people who lived largely on a diet of polished rice and was due to a deficiency of vitamin B_1 (thiamin). Rice contains plenty of thiamin but it is all removed by the polishing.

Of the fat-soluble vitamins, A, D, and E, the first two can be made by the man-body. Vitamin A can be made from carotene, the substance that gives color to the carrot. Vitamin D can be made in the skin under the influence of sunlight. A deficiency of D results in rickets (a bending of the bones in children) because vitamin D is necessary for the deposition of calcium in the bones. The best source of vitamin D is milk, which provides the calcium

WHAT CAN GO WRONG   199

as well. Vitamin E (alpha tocopherol) is needed for the reproductive process.

Calories and proteins. Any Whole-earther planning to grow his own food should have some idea what the human body consumes in a year, which is one growing cycle. Figures for an average American (civilian) have been provided by the Department of Agriculture. What are the main items? In 1970 our average American guzzled 186.2 pounds of meat, mainly beef (in the same period the average Japanese ate 20 pounds). He ate 319 eggs, 264 pounds of milk, 118 pounds of potatoes, 110 pounds of wheat flour, 102 pounds of sugar, 99.5 pounds of fresh vegetables, 82 pounds of fresh fruits. He ate only 11.4 pounds of fish. As for the poor man's food (dried beans) he only ate 5.9 pounds.

A true Whole-earther will hardly wish to follow this pattern. With its huge meat intake (not the world's highest; the prize goes to the New Zealander, who eats 220 pounds per year), the American diet is wasteful and needlessly expensive. It has been calculated that an active man could remain healthy on a daily diet of 7 pounds of potatoes and a glass of milk. The Chinese coolie with a diet of rice, Chinese cabbage and soybeans operates on an even more modest fuel intake. The male body at the height of its vigor, years 18–35, needs 2,800 calories per day containing 2.3 ounces of protein. This works out at 52 pounds of protein a year. Soybeans are about 38% protein, so a whole year's protein supply is contained, theoretically at least, in about 140 pounds of beans. The beans actually have more protein than beef though it is not quite such good quality (low in methionine). But a good high-lysine strain of corn will correct this. All man's protein needs can be supplied by vegetables without his touching meat. As a steer must eat about 10 pounds of corn to make 1 pound of beef the saving in food is considerable. But if you want to live well on a vegetable diet you must know what you are doing. If in doubt, add to your diet plenty of milk, especially if pregnant or growing. Milk contains enough of everything, the complete food. Dried milk is especially valuable. Fresh milk is 87% water. A good source of information on protein derived from plants is *Diet for a Small Planet,* by Frances Moore Lappé, Ballantine Books, 1972.

It is impossible to formulate a perfect diet because food intake depends on the work a man is doing. A logger working in low temperatures in the north woods or an oil rigger working on the north slope in Alaska may require as much as 4,000 calories a day, much of it in the form of fat. A large amount of this fuel will be consumed simply to keep the body warm. But if a sedentary office worker in an air-conditioned building starts stuffing himself with this sort of diet he will obviously become grossly overweight, and probably suffer from *atherosclerosis* as a result of his huge intake of fat.

The only sensible rule is to eat according to the *appestat.* This appestat is a center in the hypothalamus that tells one when the body needs more fuel. The rule of a healthy body is simple enough. If you aren't hungry, don't eat. When you do eat, don't stuff till you feel bloated. Those who are overweight can lose weight safely by eating a diet containing between 1,500 and 1,800 calories a day.

I have used the diet shown here, more or less, for several years, substituting beans for fish at intervals as a source of protein. I have given up bread and baked stuff, except sunflower seed cookies, to avoid using the oven. The waffle made with eggs and yeast (see Food Storing and Processing, section 8) is the largest single calorie item. I buy coffee, sugar, dried milk, orange juice, peanut butter and sunflower seeds. This diet works out at around 1,700 calories, which seems to keep a sixty-year-old body healthy and disease-free. But when I take my kayak out on the ocean and spend a day fishing, my calorie needs go up. On such days I eat maybe 3,000.

The true Whole-earther who wants to keep his arteries clear and his weight normal will avoid greasy foods, oils and fats, unless his environment is such that he must burn calories to stay warm. The Eskimos have lived largely on blubber for centuries but don't suffer from hardening of the arteries. They burn up the fat to keep warm. If someone tried living on an Eskimo diet in the trop-

Morning		Noon		Evening	
Waffle	215	Beans (lima)	100	Fish	120
Peanut butter	100	Potatoes, mashed	60	Potato	60
Blackberry jelly	60	Salad	60	Squash	50
Applesauce	150	Raw apple	100	Tomato juice	50
Coffee and milk	50	Cookies	280	Applesauce	150
Orange juice	50	Coffee and milk	50		
	625		650		430

ics the result would probably be disastrous. The lesson is, adjust your calories according to your climate and always listen to the appestat. If you keep your body on the lean side it will tell you loud and clear when to eat and about how much.

2. Intake of poisons

Various harmful substances are taken in by the body, sometimes intentionally and sometimes not. Unintentional absorbing of poisons occurs through the lungs when we breathe polluted air and through the guts when we eat contaminated food. We can also absorb poisonous impressions which have a deleterious effect on the psyche. The true Whole-earther, however, can avoid most of these poisons by living in a clean environment, drinking uncontaminated water, eating food that has not been sprayed with poisons.

Accidental poisoning will be dealt with under the heading of accidents. Here we are concerned with the intentional intake of substances harmful to the body. The commonest of these is the smoke of tobacco.

Tobacco. It is probable that no true Whole-earther would insult his lungs by contaminating the air he breathes with the breakdown products of a highly poisonous plant. His instincts would warn him that this is not a sensible thing to do. But because modern man is often not on speaking terms with his instincts and is, moreover, subjected to a barrage of advertisement from the tobacco industry, it is necessary to state clearly just what this poison does to the body.

Information published by impartial government committees in England, Scandinavia, France, Italy and the United States leads to the following conclusions: (a) Cigarette smoking is a direct cause of lung cancer. It is more likely to cause it in men than women but women also suffer. Those who indulge heavily in smoking cigarettes have a 20 times greater chance of developing lung cancer than non-smokers. (b) Cigarette smoking has been declared the most important cause in the United States of inflammation of the air tubes leading to the lungs. It also increases the chance of the smoker's developing emphysema (a breakdown of the air sacs in the lungs). Breathlessness and chronic cough are characteristic of cigarette smokers. (c) There is a definite association between cigarette smoking and heart disease, ulcers and diseased liver. Among heavy smokers death rates from coronary heart disease are 50 to 150% higher than among non-smokers. (d) Cigarette smoking is a form of air pollution which not only taints the air for the smokers but also poisons it for others. The habit is particularly obnoxious when indulged in in an enclosed space. (e) The habit adds enormously to fire hazards both in the home and outside it.

In short, there is absolutely no excuse for indulging in this disgusting habit, which, in addition to damaging the smoker's health, also damages his budget. Money spent on cigarettes could be better spent on food, clothes, books, records, etc.

Alcohol. Alcohol is not really a poison. In this respect it differs from tobacco. If taken in decent moderation and in reasonably dilute form (beer or unfortified wine) it does no harm to the body. This is because the liver contains enzymes necessary for using alcohol. These enzymes convert alcohol to acetic acid, which is used as fuel. A glass of beer (6% alcohol) provides 120 calories.

The above remarks apply only to *ethyl alcohol,* which is normally the only alcohol which occurs in fermented drinks. *Methyl alcohol* is a serious poison and the effects of drinking it are disastrous. Ethyl alcohol becomes a poison when taken in excessive amounts. The liver has only a limited capacity to burn alcohol. When excessive amounts are taken, the liver becomes strained by its efforts to remove the alcohol from the bloodstream.

Alcohol is probably the most abused drug in the world. It is abused because people come to rely on it as a means of relieving tension and finally become totally dependent on its effect. They become alcoholics. A persistent and excessive intake of alcohol has the following effects on the body: (a) chronic irritation of the stomach and intestine, resulting in nausea, vomiting and diarrhea, (b) reduced food intake partly on account of a and partly on account of the fact that alcohol provides calories, (c) liver damage due partly to direct effects of alcohol and partly to reduced food intake, (d) damage to the nervous system resulting in *delirium tremens,* which is also due to bad diet and above all to lack of B vitamins.

The true Whole-earther will know, if his instincts work properly, just how much of his own home brews he can sample without endangering his health. It is not difficult to know when one has had enough. As soon as the alcohol taken begins to interfere with one's muscular coordination it is time to stop drinking and let the body clear itself. The rate at which the body burns its alcohol depends somewhat on circumstances. Beer containing about 6% alcohol can be consumed in astonishing amounts by a man sweating in the harvest field without its having any noticeable effect. If he drinks these same amounts sitting at the dinner table he will probably become intoxicated. Wine, which contains about 12% alcohol, puts more of a strain on the body's detoxifying machinery than beer. Whisky, gin, vodka and other distillates contain still higher concentrations of alcohol. These have a harmful effect on the lining of the stomach and should therefore not be taken on an empty stomach and preferably should be avoided altogether.

Drugs. The word drug in our culture is used so loosely that it has practically no meaning. All sorts of substances that have practically nothing in common are lumped together and called drugs. The word inspires vague terrors among uninformed people who still think in terms of "dope fiends" and hardly know the difference between heroin and marihuana. As for the true Whole-earther, he is not very likely to bother with drugs because he has no need to escape reality and is not exposed to the temptations that assail the city dweller.

The facts about drugs are fairly simple. They fall into three classes: (a) substances that alter one's perception of the world, such as marihuana, LSD, mescaline, psilocybin, (b) substances that pep you up, such as amphetamines (speed), cocaine, caffeine, (c) substances that calm you down, such as heroin, barbiturates, various tranquilizers. They are all undesirable because they interfere with the individual's inner development. You cannot grow inwardly if you are a slave to a drug. Heroin addiction in the United States is simply an expression of a spiritual malaise brought on by the stresses of an urban environment or a general urge to escape reality.

As far as their poisonous effects are concerned, amphetamines and cocaine are the most dangerous; heroin will enslave you and destroy your ambition; barbiturates, taken to excess, produce a worse form of physical dependence than heroin. Marihuana, smoked to excess, is probably as damaging to the lungs as tobacco, but is otherwise fairly harmless. The natural psychedelics, peyote and psilocybin, may make you sick. The "devil's weed" (*Datura inoxia* or *Atropa belladonna*) is the main ingredient of the old-fashioned witches' brew and apt to be lethal if imbibed too freely. *Ayahuasca* and *yage* seem to be fairly safe but you must go to the forests of the Amazon to get them. These natural psychedelics *may* give one interesting insights but, after experimenting with most of them, I can say they are no substitute for disciplined inner work in the form of yoga, zazen or Sufi exercises. They may prove useful if taken once or twice but frequent indulgence only results in weakness.

3. Infectious diseases, prevention

The right attitude toward infectious diseases is summed up in the old saying, "Prevention is better than cure." All infectious diseases are caused by parasites, which may be viruses, bacteria, protozoa, worms or flukes. All these parasites are carried by contaminated material, air, food, water, etc., which introduce the disease germs into the body. The first step, therefore, in preventing infectious disease is to avoid contact with contaminated materials. This may prove difficult but certain elementary precautions can be taken.

Organisms of disease are carried in the following ways:

Air contamination. Most of the parasites which attack the respiratory passages (nose, throat, lungs) are spread by air. A sneeze is a powerful explosion expelling particles at velocities up to 100 miles an hour, so an unmuffled sneeze will contaminate a whole roomful of air. The commonest air-borne parasites are the viruses of common cold, influenza and measles, the bacteria of tuberculosis, whooping cough, scarlet fever and diphtheria. Be considerate. If infected, wear a face mask or sneeze into a handkerchief.

Water contamination. Water can carry a multitude of parasites, including the following: viruses of poliomyelitis and hepatitis, bacteria of cholera, dysentery, typhoid fever, also various parasitic worms, the cercariae of schistosomiasis, the amebas of amebic dysentery. Water becomes contaminated by contact with human urine or feces, either because raw sewage mixes with the water or because contaminated human bodies are washed in the water. Sewage contamination can be avoided by keeping at least 50 feet of soil between privy and water supply. Avoid careless bathing. An entire hippy community in New Mexico was infected with amebic dysentery because a carrier bathed in a hot spring used by the community. If you have doubts about your water supply, either boil it or chlorinate it but, better still, take steps to prevent contamination.

Soil contamination. Almost all of the organisms that can contaminate water can also contaminate soil but some are particularly dangerous. The organism of amebic dysentery is one of the worst soil contaminants and vegetables which are eaten raw, such as lettuce, may easily carry this parasite if grown on infected soil. This is why human shit, if used as manure, must be properly composted before being spread on the soil. Soil contains, in addition, the organisms of tetanus and botulinus. Tetanus bacteria grow in wounds contaminated with dirt. Botulinus bacteria grow in soil-contaminated food. Both produce toxins that are among the most deadly substances known to man. Any dirty, soil-infected wound should be carefully opened up and cleansed. An injection of antitetanus serum is a worthwhile precaution. Tetanus (lockjaw) is not an agreeable death.

Food contamination. Food may become contaminated by contact with contaminated soil or water. More frequently it is contaminated by the hands of the person preparing it. The celebrated "Typhoid Mary" spread typhoid far and wide because she worked as a cook, she did not wash her hands properly after defecating, she was a carrier of the disease, which means she had the bacteria in her gut without symptoms of illness. Moral: always wash carefully after defecating, especially if you are going to handle food. Food can also be contaminated with the organism of botulism from the soil but this organism produces its deadly toxin only in the absence of air. It is typically a contaminant of canned foods. The gas it produces causes the can to bulge. Always throw out cans of food that bulge.

Food can also be contaminated by bacteria called salmonellae and staphylococci, both of which cause food poisoning, with vomiting, abdominal pain and diarrhea. Salmonellae grow on such food as ham, potatoes, fish, and shrimp or chicken salad if they are kept too long and not refrigerated. The bacteria get into the intestine, multiply and cause the symptoms. Travelers in foreign lands are often laid low by salmonellae that have no effect on the local population. Staphylococci grow in such foods as cream puffs or eclairs and generate a toxin that causes the symptoms. So avoid keeping custard pastries too long in hot weather.

Three other dangerous parasites can be carried by food. Brucellosis (Bang's disease of cattle) causes undulant fever in man. Bovine tuberculosis can cause tuberculosis in man. Both are milk-borne and can be eliminated by pasteurizing milk or by testing the animals to make sure that they are disease-free. Trichinosis is caused by a parasitic worm that may be present in pork. Always cook pork thoroughly to destroy these parasites.

Insect contamination. Insects which are themselves parasites on man may become contaminated with a variety of disease organisms. Bubonic plague is carried by fleas from rats to man. Malaria and yellow fever are carried by mosquitoes, typhus by lice, sleeping sickness by the tsetse fly, and Rocky Mountain spotted fever by ticks. The very damaging disease schistosomiasis (also called bilharziasis) is carried by water snails, which in turn contaminate the water in which they live. All these infections can be avoided by avoiding contact with the insects or, in the case of schistosomiasis, avoiding contact with contaminated water.

In addition, the ever present housefly can carry disease and contaminate food if it has the chance to crawl on human feces, hence the importance of fly-proof privies.

Sexual contamination. The organisms of two infectious diseases, gonorrhea and syphilis, are contaminants of the sex organs, male or female, and are spread by sexual intercourse. These diseases are both on the increase due to a combination of increased promiscuity and carelessness. Both diseases can be avoided by proper care in sexual contacts.

4. Infectious diseases, immunization

Prevention of infectious disease by avoiding contact with the parasites is certainly the perfect method but not always practical. Children, in particular, are exposed every time they go to school to a variety of infections ranging from colds to measles. Fortunately, it is possible to increase the resistance of the body to many infectious diseases by exposing it to the organism of the disease in an attenuated form, through immunization. There are some parents who view all artificial immunization with suspicion or deny their children protection for religious reasons. Such denial represents gross negligence or defective thinking on the part of the parent. If immunity to a disease can be purchased at the cost of a few dollars and a needle prick it is criminal to leave a child unprotected. There are many products of technology we might be better off without, but vaccines are not one of them.

The immunization schedule shown here is recommended for children by the American Academy of Pediatrics.

Age	Protective agent
Any time after 6 weeks	Polio vaccine, injection or oral
1–2 months	First of triple vaccine injections: diphtheria, tetanus, pertussis (whooping cough)
2–3 months	Second triple vaccine injection
3–4 months	Third triple vaccine injection
5–6 months	Smallpox vaccination
16–18 months and about 4 years	Booster shots of triple vaccine
6 years or on entering school	Second smallpox vaccination

Many other vaccines are available and should be used if a child or adult is likely to be exposed to the disease in question. Measles and mumps, for instance, are not generally taken very seriously but measles can have serious aftereffects and mumps in the adult male can result in sterility, so artificial immunization should be considered. German measles in a pregnant woman is a threat to the unborn child. Dwellers in the United States are not likely to encounter typhoid, paratyphoid, cholera, yellow fever or typhus, but as soon as they travel to areas where public health measures are below minimum standards they will be exposed to these diseases and should therefore be immunized.

Rabies (hydrophobia) is another disease to which few people are likely to be exposed. It is such a terrible disease, however, that anyone, child or adult, who has been bitten by an animal that might be rabid should receive a course of antirabies injections. Here in California the commonest rabid animals are skunks, which become seemingly tame when infected with the disease. Children should be warned not to touch such tame skunks.

Vaccines against various forms of the influenza virus are available and may be worth using if a virulent influenza epidemic is under way.

INTERPRETING SIGNS AND SYMPTOMS

When the inner harmony of the body breaks down, the result is a set of inner sensations felt by the body's owner (signs) and a set of outer manifestations that the physician can observe (symptoms). The interpretation of signs and symptoms with the help of various laboratory procedures is the process called *diagnosis*. As correct treatment depends on correct diagnosis this part of the physician's work is very important.

The Whole-earther who may be living far from medical aid needs to be familiar with some of the more obvious signs and symptoms of disease. These can be conveniently divided as follows:

1. Gastrointestinal disturbances

(a) *Gastric influenza*. Vomiting, diarrhea or both are the symptoms. The disturbance is caused by a group of viruses called enteroviruses. Acute attack rarely lasts longer than a few hours. No need to call a doctor, no curative agent is known. Rest and take plenty of fluids.

(b) *Food poisoning*. Vomiting and diarrhea occurring in several people who have eaten the same food suggest food poisoning. In the staphylococcal type, symptoms show within a few hours of eating poisoned food. In the salmonellae type, onset of symptoms is not so rapid. They take 6 to 24 hours to develop and may persist over several days. Not much use calling a doctor. There is little he can do. Rest, and take fluids in small amounts until the intestine settles down.

(c) *Bacillary dysentery*. The mild form (Sonne dysentery) may occur in children. Predominant signs are diarrhea, blood and mucus in the stools. The number of stools per day rarely exceeds 5 or 6. More virulent types (Flexner or Shiga) are rarely seen outside the tropics. In these types as many as 20 stools per day may be passed. The typical dysenteric stool is small and consists entirely of blood and mucus. Get medical help. This is a serious, highly infectious disease and hospitalization is desirable. Sulfonamides (sulfaguanidine) or antibiotics (tetracycline) help eliminate the infection.

(d) *Amebic dysentery.* The amebic parasite produces inflammation of the colon, with severe diarrhea and blood and mucus in the stools. For certain diagnosis, microscopic examination of the stool is necessary. Get medical help. This disease can become chronic and can damage the liver, so drastic steps to eliminate the amebas may be called for. Injections of emetine are used for this purpose but the drug is very toxic and should be used only with medical supervision.

(e) *Cholera.* This severe, often fatal disease is nowadays seen mainly in the tropics. Intense diarrhea with watery stools containing shreds of mucous membrane from the intestinal wall is accompanied by severe dehydration. Relief can be obtained by intravenous replacement of body water and salts. For this major medical emergency, get help.

(f) *Summer diarrhea.* This illness occurs in infants and was once a major cause of infant deaths. Various things can upset the digestive system of an infant on artificial food and give rise to colic, vomiting and diarrhea (such as too strong a milk mixture or excessive use of sugar in the feeds). Infectious summer diarrhea is prevalent when the weather is hot and it is due to microorganisms in the milk. The infant vomits its feed. Diarrhea sets in quickly and the stools are frequent, watery and green. If the infant becomes lethargic and signs of dehydration appear (eyes become sunken, skin when pinched remains in a fold), medical aid should be urgently sought. Otherwise keep the infant warm, give only fluids (boiled water, glucose water) for 12 to 36 hours, depending on persistence of vomiting. Total amount of fluid per 24 hours can be calculated on the basis of 2½ ounces fluid per pound body weight plus 5–15 ounces necessary to overcome existing dehydration. Once vomiting stops, milk feeds can be started again, milk being diluted with three or four parts water. Only infants who are artificially fed will be in danger of diarrhea. Mother's milk is best.

(g) *Ulcerative colitis.* Chronic diarrhea with watery stools containing blood and mucus is the main symptom of this disorder. It is not due to any infection of the lower bowel but seems to be related to anxiety and feelings of guilt. Get medical help. It is essential to rule out cancer of the lower bowel. The illness tends to be chronic. Its ravages can be reduced by rest, absence of stress, and a low-residue diet (milk, boiled or poached eggs, mashed potatoes, custards, steamed fish, fruit juices).

(h) *Ulcers of stomach or duodenum.* Pain above the stomach region comes on either before meal or just after one. Vomiting often occurs if pain is severe. Symptoms come on in attacks which may last for several weeks and then disappear, only to recur later in a more aggravated form. This breakdown of the body's inner harmony can usually be attributed to stress. The ulcer is due to overproduction of acid in the stomach and cure of the ulcer depends on rest, frequent feeding (every 2 hours) with milk and other non-irritating foods, alkalis such as magnesium trisilicate to neutralize acids, and drugs which relieve muscle spasm caused by the ulcer. Get medical help. Diagnosis is almost impossible without x-rays and laboratory tests. Acute pain in the upper abdomen may be caused by inflammation of the gall bladder. (The pain in this case often radiates to the tip of the right shoulder.) Vomiting with blood may be a symptom of stomach cancer. It takes an experienced physician to make the right diagnosis.

(i) *Constipation.* This condition is usually due to bad habits. The human bowel is a highly rhythmic affair and the owner of that bowel is well advised to heed those rhythms if he wishes to enjoy perfect health. After a meal a reflex action calls for evacuation of the bowel. This gastrocolic reflex occurs in most people after breakfast. If they refuse to heed this call from their instinctive center the feces become dry and hard from loss of water and therefore more difficult to pass. The condition can be remedied by establishing regular habits and increasing the amount of roughage in the diet with green vegetables, cereals, bran, fruits. Purgatives are never desirable. They irritate the rest of the intestinal tract and can be habit-forming. A warm water enema is the safest and surest way of getting rid of accumulated feces. Very severe constipation is a danger sign and may indicate an obstruction of the bowel. Get medical help.

(j) *Acute appendicitis.* This is a common cause of acute illness involving nausea, vomiting and abdominal pain. The pain starts in the pit of the stomach, gradually shifts to the lower right side. There is acute tenderness and rigidity of muscles in the lower right side of the abdomen. The victim of appendicitis must be taken to the hospital in a hurry. Delay may result in a ruptured appendix and peritonitis. Do not give victim a laxative or any medication by mouth. He should be placed in a semi-reclining position, a pillow under his knees to keep them flexed, an icebag wrapped in a towel applied to the lower right side of the abdomen.

(k) *Intestinal worms.* Three forms of intestinal worms are apt to occur in temperate regions. Pinworms, or threadworms, are common in children. The chief symptom is itching in the anal region. The tiny white worms can be seen in the feces or around the anus. Children reinfect themselves by scratching the anus and transferring the worm eggs to their mouths. Eradication of the worms is possible by use of piperazine adipate and gentian violet capsules. Roundworms gain access to the intestine from contaminated raw vegetables or water. They appear in the stools. They can be eliminated by means of hexylresorcinol tablets. Tapeworms can be acquired from either beef or pork that has been inadequately cooked. These worms may attain a length of several feet. They are made up of a head and a long string of segments which drop off in turn and can be seen in the feces. Symptoms are slight: excessive appetite, mild abdominal colic and perhaps loss of weight. The worm is hard to eliminate and special treatment is required.

2. Respiratory disturbances

(a) *Colds and influenza.* These common virus infections inflame the mucous membrane lining of the nose and throat. A copious flow of watery secretion results, accompanied by coughing and sneezing. The only remedy for this condition is warmth, rest and plenty of fluid.

(b) *Sore throat and tonsillitis.* The commonest cause of sore throat is a streptococcal infection of the tonsils. When the back of the throat is examined it is seen to be red, and usually white spots are present on the tonsils. The infection can be treated with penicillin or sulfonamides, but before these drugs are used a test must be made to determine whether the streptococcus really is present. Although there is absolutely no reason to fear diphtheria, provided a child has been properly immunized, this dangerous throat infection can occur in non-immune children. A severe sore throat, especially in children, should be taken seriously, especially if there is high fever. Get medical advice.

(c) *Hay fever.* This illness is due to allergy, generally to some form of pollen (ragweed, grass, etc.). During months when these pollens are in the air the sufferer is prone to paroxysmal attacks of sneezing associated with running nose and eyes. Various drugs called antihistamines will reduce these symptoms. Some are more potent than others. They tend to induce drowsiness and should therefore be used with caution. Desensitization is possible by injection, providing the identity of the offending substance is known.

(d) *Bronchitis.* The bronchi are prone to infection by streptococci, pneumococci and the virus of influenza. A cough is the commonest symptom, associated with mucoid or purulent sputum. Mild fever, headache, loss of appetite will also occur. The infection generally subsides fairly rapidly, provided the patient is kept warm, rested, given plenty of fluids. The condition must be watched carefully in children and elderly people as the inflammation may spread and cause bronchopneumonia.

Repeated attacks of acute bronchitis aided by a damp climate, and exposure to dust, may result in chronic bronchitis. The sufferer has what is loosely known as a weak chest, complains of a persistent cough, which is generally worse in the winter. This condition, often greatly aggravated by smoking, commonly leads to *emphysema*. This is a distention and thinning of the lung tissue which prevents expulsion of used air. The sufferer has increasing difficulty in getting enough oxygen into the blood. This puts strain on the heart, which must work harder to circulate what oxygen there is. There is little one can do to cure emphysema. A move to a warm, dry climate may result in improvement.

(e) *Pneumonia*. In this condition the alveoli (air sacs) of the lungs become infected. The onset is sudden with a high fever (102° to 104° F), shivering attacks, and severe pain in the chest caused by inflammation of the linings of the lungs (pleura). The face is flushed, respiration is rapid, breathing is distressed, and a short, suppressed cough is present. After a day the cough becomes moist with typical rusty sputum. Get medical help. If caused by a bacteria the infection can generally be controlled by sulfonamides or penicillin, but other common types of pneumonia, caused by a virus or a mycoplasm, are more difficult to treat. For this reason pneumonia must be brought to a doctor's attention.

(f) *Pulmonary tuberculosis*. There are several forms of tuberculosis but the commonest form affects the lung. It is contracted by inhaling droplets sneezed or coughed into the air by one infected with the disease. The earliest symptom is a cough, at first dry but later moist with sputum. There is a progressive loss of weight associated with general fatigue. The temperature rises in the evening and there are heavy night sweats. Any person, especially a young adult, who is losing weight, has a persistent cough, especially if there is any history of spitting up blood, should be suspected of having tuberculosis of the lung and given an x-ray examination. Streptomycin and isoniazid made this illness easier to treat than it once was but it is still a very serious disease and special steps must be taken to prevent the spread of infection. Get medical aid.

(g) *Lung cancer*. This increasingly common form of cancer shows many of the same symptoms as tuberculosis of the lung. It is associated with heavy cigarette smoking, occurs most often in middle-aged or elderly people, affects men more often than women. A persistent cough associated with sputum which may be blood-stained, plus loss of weight, energy and appetite is about all one can observe in the early stages. In most cases by the time the cancer has produced enough signs and symptoms to establish diagnosis the growth has spread too far to be removable.

(h) *Asthma.* This is an allergic disease similar to hay fever except that it affects the air sacs of the lung. The cause is generally hypersensitivity to some foreign protein. A sudden attack of breathlessness is the chief symptom. Breathing becomes more difficult, until respirations are labored and wheezy. Various drugs (ephedrine, isoprenaline or aminophylline) will reduce the severity of the attack. In some cases the sufferer can find out what substances cause the attacks and have himself desensitized.

3. Circulatory disorders

In the United States more than 500,000 people suffer "heart attacks" every year. It is one of the commonest forms of death. The cause of these heart problems is a partial or complete blocking of the coronary arteries that supply the heart with blood. The heart is a muscle and needs a constant supply of blood in order to function.

(a) *Angina pectoris.* This pain in the chest occurs when an insufficient blood supply reaches the heart. The attack follows exertion, emotional strain or a heavy meal, anything, in short, that increases the work of the heart. The attacks are characterized by the following symptoms: pain under the breastbone radiating down the left arm or both arms, sensation and fear of impending death, a feeling of suffocation, cold sweating, skin color ashen gray or bluish, dizziness, unconsciousness.

Attacks of angina pectoris normally do not last more than a few minutes and are relieved by rest. The attacks are a warning sign. They indicate a defect in a very vital organ and tell the owner of that organ to be careful. Physical exertion intense enough to bring on pain should be avoided. Overexcitement and emotional upsets can also be harmful. Meals should be light and excess weight should be removed by dieting. Physicians commonly prescribe drugs of the amyl nitrite type that dilate the coronary artery.

(b) *Heart attack.* Coronary thrombosis is more serious than angina pectoris. In this condition the already narrow coronary artery is completely blocked by a blood clot. Part of the heart muscle is completely deprived of oxygen. Sudden intense pain results. It is located in the upper part of the chest, lasts much longer than angina pectoris and is not related to effort. Coronary thrombosis is a major emergency. Get medical aid fast.

(c) *High blood pressure.* Hypertension is one of the commonest forms of illness in the United States. The blood that flows in our arteries exerts a pressure that changes all the time. When the heart contracts, the pressure goes up (*systolic* pressure). When the heart pauses to fill, the pressure falls (*diastolic* pressure). The pressure is measured in millimeters of mercury. It varies with age. Normal range for people of forty to sixty years of age is 110–140 systolic, 60–90 diastolic (written 140/90). Readings of 160/100 are not generally considered dangerous but when systolic pressures go up to 200 symptoms are likely to be experienced. These are headache, giddiness, ringing in the ears. Treatment consists of avoiding tension, worry and emotional upsets. If obesity is present, weight reduction is desirable. A low-calorie, low-salt diet will help. Various drugs also can be used to lower blood pressure but because they are potentially toxic should be taken only under medical supervision.

(d) *Congestive heart failure.* The condition is due to failure of the heart to do its work of pumping blood from the veins through the lungs. Blood stagnates in the veins and congestion results. Symptoms include breathlessness, a dark purplish coloration of the skin and mucous membranes due to insufficient oxygen in the blood (cyanosis), swelling, especially of the legs, due to accumulation of fluid (edema), highly concentrated, dark-colored urine due to congestion of the kidneys. Treatment is complex and specialized, involving salt-free diet, diuretics, oxygen and digitalis. Get medical aid.

(e) *Congenital heart disease.* The symptoms are similar to those of congestive heart failure but they occur in children. The cause is a malformation of the heart, which causes venous blood to mix with arterial blood. The child has a characteristic blue look (blue baby), is breathless, and the ends of the fingers and toes are often enlarged. The condition may be corrected by an operation. Get medical aid.

(f) *Rheumatic heart disease.* This is one of the results of rheumatic fever, a disease associated with streptococcal infection in children and young adults. The illness generally starts as a sore throat, followed by general malaise with high temperature and heavy sweating. Pains occur in the joints. The pulse is fast and may be irregular. The disease may permanently damage the valves of the heart, for which reason expert treatment is desirable.

(g) *Arterial embolism.* Blood clots can lodge in the arteries, closing the blood vessel. If the clot lodges in a brain artery the result is a "stroke," of which the main symptom is paralysis of one side of the body. If the clot lodges in the artery of the leg severe pain is felt in the limb, which becomes white, cold, paralyzed. Gangrene (rotting of the flesh) sets in unless the blockage can be removed. Arterial embolism is a major emergency. Get medical help.

(h) *Varicose veins.* This is a widespread condition and one of the prices man pays for having learned to stand upright. The surface veins in the legs become swollen, especially on standing. The condition is especially common in women during pregnancy. Exercising the legs, avoiding long periods of standing, supporting the veins with stockings or bandages all help to prevent the condition from getting worse. Surgical treatment may be employed but should be regarded only as a last resort. Though varicose veins never get better, intelligent care will prevent them from getting worse.

4. Nervous system disorders

Disorders of the nervous system interfere with the transmission of messages to the muscles or with the transmission of sensations to the brain. These disorders are quite different from the group of diseases generally lumped together as "mental illness." In mental illness, which includes schizophrenia, paranoia and various forms of neurosis, there is nothing wrong with the nerves. The defect may lie in the chemistry of the brain, but this defect remains elusive. Despite intensive research the causes of mental illness remain obscure. This is not at all the case with disorders of the nervous system. Here the symptoms are dramatic and the cause can generally be defined.

(a) *Apoplexy (stroke)*. This is a very common disease. The main symptom is paralysis either of one side of the body (*hemiplegia*) or of both sides (*paraplegia*). The cause may be a blood clot which has blocked a blood vessel in the brain or the rupture of an artery in the brain. If the damage occurs on the left side of the brain the speech center may be affected and speech will be disturbed. In a mild stroke paralysis may occur without loss of consciousness, but in most cases the patient lapses into coma. Breathing becomes very deep and stertorous, the pupils are dilated, there is incontinence of urine and feces.

Unconsciousness is not, in itself, a sure sign of a stroke. A person may become unconscious as a result of heat stroke, head injury, epilepsy, too much or too little blood sugar (diabetes), heart attack, drunkenness. But coma with paralysis indicates apoplexy. Keep the victim warm, with the head on one side to prevent obstruction of the breathing passages. If breathing has ceased, try mouth-to-mouth resuscitation. Get medical aid.

(b) *Meningitis*. This disease results from an infection of the meninges, the membranes covering the brain. The signs and symptoms include headache, which comes on early in the disease and is severe. The headache is associated with vomiting. There is a high fever. The patient is drowsy, irritable and often delirious. Fits are common, especially in infants. There is marked neck rigidity and resistance to any attempt to straighten the flexed knee. Meningitis is a major emergency and medical aid must be urgently sought. The disease was once usually fatal but the introduction of sulfonamides such as sulfadiazine and antibiotics has completely changed the outlook. The vast majority of patients now recover.

(c) *Convulsions (fits)*. Convulsions are severe, uncontrollable muscular spasms. They are quite commonly seen in infants or children. Any severe general illness or fever in an infant may begin with a convulsion. Whooping cough, measles, pneumonia, gastrointestinal disturbances and intestinal worms can cause convulsions.

Convulsions in adults can have a variety of causes such as severe head injury, renal failure, severe high blood pressure, toxemia of pregnancy (*eclampsia*) and tumor of the brain.

In all these conditions the convulsions are indications of a disorder which, in many cases, is not located in the nervous system at all. In *epilepsy*, however, the convulsions are directly related to damage or malfunctioning of the brain. Epilepsy usually begins in childhood. It takes two main forms. In major epilepsy the patient falls unconscious, all the muscles go rigid, breathing ceases and the face goes blue. Next, spasms of the muscles occur with violent movements of the limbs. Frothing at the mouth may occur, along with incontinence of urine and feces. Later the patient generally passes into a coma, which changes into deep ordinary sleep. In minor epilepsy the attacks are briefer. The fit consists of a transient loss of consciousness, which may be so brief that the onlooker notices nothing. The patient may describe this attack as a "blackout."

A person in an epileptic fit must first be prevented from injuring himself. Tight clothing should be loosened, a gag of some sort put in the mouth to prevent biting of the tongue. The fit seldom lasts longer than 2 minutes. Most epilepsy nowadays can be controlled by such drugs as phenobarbital and dilantin.

(d) *Parkinsonism (paralysis agitans)*. The chief symptom of this disorder is tremor. The hand shakes and there is constant movement of the fingers. Movements are rigid and the face tends to be mask-like. The condition generally occurs late in life and is due to a defect in the chemistry of the brain. It is treated with massive doses of a substance called L-dopa.

(e) *Mental deficiency*. This may result from brain injury at birth, from a chromosome defect (mongolian idiocy), from a thyroid deficiency (cretinism). Brain injury at birth frequently results in cerebral palsy, the chief symptom of which is muscular tension (spasticity). Often the legs are so rigid and spastic that the child has great difficulty in walking. Cerebral palsy may affect the musculature only, leaving the intellectual part of the brain undamaged. Mongolism is commonest in children born of mothers over age forty-five. The condition is incurable. Mental deficiency of the child can also result from an attack of German measles (rubella) during pregnancy. The problem of parents with a mentally deficient child is whether to keep it in the home or place it in an institution. Our civilization is needlessly sentimental about these damaged beings and many children are kept alive at great cost who would probably be better off dead.

(f) *Sciatica*. This common ailment is not really a disease of the nervous system but another result of man's habit of walking upright. The spinal column is made up of separate vertebrae, each of which is cushioned by a small disc that acts as a shock absorber. Damage to the lower back can result in displacement of the disc, which presses on the root of the sciatic nerve. Pain is felt in the back and down the back of the thigh to the ankle. Treatment consists in rest. The patient is kept lying flat until the pain is gone, after which he can gradually resume activity. A belt may be needed to give adequate support to the back. If the sciatica recurs, surgical removal of the displaced disc may be necessary.

5. Liver disorders

The old joke, "Is life worth living? That depends on the liver," is more than a play on words. A malfunctioning liver can certainly take the joy out of life. Anyone who lives reasonably, avoids such poisons as tobacco and uses alcohol only in moderation can trust his liver to function quietly and stay well below the level of consciousness. However, there are two fairly common forms of liver disorder, jaundice and hepatitis, which may affect anyone.

(a) *Jaundice*. Jaundice refers to a yellow discoloration of the skin and whites of the eyes. It is caused by the presence of yellow bile pigment in the blood. Bile is produced by the liver and plays an important part in the digestion of fats. It is stored in the gall bladder and passes into the small intestine via a duct which it shares with the pancreas. The duct may become obstructed by gall stones, in which event the bile instead of passing into the bowel is absorbed by the blood. Jaundice is the most obvious symptom. There will also be itching of the skin, dark brown urine and pale feces (the normal dark brown color of feces is due to bile pigment). Pain is felt in the region of the lower ribs and radiates to the tips of the right shoulder. This condition calls for surgical aid, generally removal of the gall bladder.

(b) *Infectious hepatitis*. This liver disease is caused by a virus. Epidemics of hepatitis which occur from time to time suggest faulty hygiene because the virus is excreted in the feces. Onset of the illness is gradual with loss of appetite and nausea. The sight of food, especially fats, is enough to cause nausea, even vomiting. There is headache and a slight fever. After a few days the urine becomes dark, the conjunctiva and skin become yellow and the feces pale.

Recovery from virus hepatitis can be hastened by rest, plenty of fluids and a high-protein diet. Special precautions should be taken to prevent spread of the virus by fecal contamination. A second form of hepatitis (serum hepatitis) is transmitted by transfusions of serum or inadequately sterilized syringes.

(c) *Alcoholic cirrhosis*. The disease is one of the results of abuse of alcohol. For reasons not understood this abuse causes obstruction of the bile ducts in the liver, also congestion of the veins in the stomach and rectum. Severe loss of appetite, vomiting of blood, and fluid in the abdominal cavity (ascites) are symptoms of the illness. In the early stages a cure can be brought about by complete abstinence from alcohol, and a good diet. In later stages liver failure results in death.

6. Urinary system disorders

The function of the urinary system is to remove waste substances formed during the working of the body. Two parts of the system can become disordered, the kidneys and the urinary bladder. The kidneys, which are very complex organs, can become deranged in the following ways:

(a) *Nephritis* (*Bright's disease*). In children and young adults this disease can take an acute form. The illness often follows a sore throat or attack of scarlet fever. Symptoms of nephritis may appear 7 or 10 days after these ailments. Fever, headache, vomiting, pain in the back are followed by edema. This puffiness of the tissues due to accumulation of fluid is especially evident around the eyes and in the ankles. The edema gives the patient a bloated appearance. There is diminished output of urine, the urine is dark and concentrated and becomes turbid on boiling, due to albumin. Such symptoms must be brought to a doctor's attention.

Most patients with acute nephritis recover completely, given rest and the right diet. Very little food, even a period of starvation, with only 2 pints a day of sweetened orange juice, gives the kidneys a chance to recover. Once the kidneys improve, the volume of urine rises and the diet can be cautiously supplemented with such items as milk, cereal, fruit and eggs.

In a few cases acute nephritis fails to clear up and the kidneys are permanently impaired. Such chronic nephritis calls for special medical treatment.

(b) *Uremia.* This word means that urea is present in large amounts in the blood and indicates serious breakdown in kidney function. Kidney failure can be acute or chronic. In acute renal failure there is complete or almost complete suppression of urine. The condition is generally due to poisoning or severe shock. Acute renal failure is a serious condition and calls for expert medical treatment.

Chronic renal failure also produces uremia. The symptoms are fatigue, exhaustion, high levels of blood urea, changes in the retina of the eye, increasing drowsiness and labored breathing. The total amount of urine is normal. Chronic renal failure can to some extent be corrected by purification of the blood in an artificial kidney machine.

(c) *Pyelitis.* Kidney inflammation caused by bacteria can happen quite suddenly, the attack being accompanied by fever, shivering and headache. Severe pain is felt on the affected side, the pain sometimes radiating downward and forward to the groin. Urination is frequent and painful. Get medical help. The infection will respond to sulfonamides but some organisms are resistant.

(d) *Cystitis*. This is an infection of the urinary bladder due generally to the same organisms as those causing pyelitis. The condition occurs more commonly in the female than the male. Pain and frequency of urination are common symptoms. The condition can be cleared up with sulfonamides or antibiotics.

(e) *Kidney stones*. Calculi, or stones, can occur in the kidneys. They may remain in the kidney and cause no symptoms or may move, perhaps blocking the tube which conducts the urine to the bladder (ureter). In this case severe pain occurs and the urine contains albumin and blood. Get medical aid. If the stone cannot be passed through the ureter it may block the flow of urine, causing swelling and damage to the kidney (*hydronephrosis*). The damaged kidney may have to be removed.

7. Bone and joint disorders

The human body contains over 200 bones, most of which slide or hinge on each other. Joints have a tendency to become inflamed (arthritis) and this inflammation is a great cause of human misery. Its cause is one of medicine's major mysteries.

(a) *Rheumatoid arthritis*. The disorder occurs more often in women than men. It is very common and can be so severe that the sufferer becomes disabled or even bedridden. The symptoms may start as fever, anemia, loss of energy and weight. Swelling and pain in the joints, particularly of the hands and feet, become the main symptoms as the disease progresses. The affected joints become thickened and their movement limited. Power of movement may be lost and the whole joint thickened and deformed.

The course of this disease depends to some extent on the will of the sufferer. Although, during some of the acute phases, bed rest may be necessary, it is the regular working of the joints that prevents or delays fixation. Heat applied by infrared lamps may be useful in relieving pain and warm water baths can make movement easier. Cortisone and corticotropin, hailed as miracle drugs when first discovered, have proven disappointing because of their side effects. Phenylbutazone is effective in relieving pain but is toxic and tends to produce anemia.

(b) *Osteoarthritis*. This is a common but less destructive form of joint disease than rheumatoid arthritis. It affects mainly people over forty and appears most commonly in the weight-bearing joints of the lower limbs, especially knees and hips. Heat treatment and active movement of the limbs will help to prevent the joints from becoming fixed. If there is obesity weight reduction is in order.

(c) *Rickets*. This disorder of the bones is due to inadequate vitamin D in the diet of an infant. The legs become bowed in the child as soon as it begins to walk. The spine may be bowed or twisted. Vitamin D plus plenty of sunshine will prevent this disease or cure it in the early stages.

8. Blood disorders

Blood disorders are of three kinds, those affecting the red cells, the white cells and the blood-clotting mechanisms.

(a) *Anemia*. This is the commonest disorder affecting the blood. If the number of red cells in the blood (normally about 5 million per milliliter) or the level of hemoglobin in these cells (normally about 14.5 grams per 100 milliliters of blood) becomes reduced the body will suffer from partial oxygen starvation. This is because oxygen is carried by hemoglobin which is, in turn, carried by the red cells. The symptoms of anemia are as follows: pallor of the skin and mucous membranes, especially of the lower eyelid; weakness, giddiness, fainting, and, in women, failure to menstruate; increased heart rate due to the fact that the heart must pump harder to get enough oxygen-deficient blood to the vital organs; breathlessness; and swelling of the ankles.

Anemia can result from loss of blood due to injury or some other cause like bleeding ulcers or bleeding after childbirth. Such cases constitute medical emergencies and must be dealt with by means of blood transfusions. Iron-deficiency anemia, both in adults and in infants, is treated with various ferrous salts in tablet form. Pernicious anemia, in which there is often soreness of the tongue, a mild degree of jaundice and a lack of hydrochloric acid in the stomach, in addition to the usual signs of anemia, is due to lack of vitamin B_{12}. Sufferers from pernicious anemia must take extra vitamin B_{12} for the rest of their lives or take large amounts of liver extract. A third kind of anemia results from excessive destruction of the red cells in the blood. This is called hemolytic anemia and is seen after certain severe infections (streptococcal septicemia, malaria), certain forms of poisoning and incompatible blood transfusions. The Rh factor in the blood can cause serious anemia in the newborn if the child is Rh-positive and the mother Rh-negative. The child is born anemic and often jaundiced and can be saved, in severe cases, only by the removal of as much of its blood as possible and replacement with Rh-negative blood.

(b) *Blood-clotting disorders.* Blood contains some rather elaborate chemical mechanisms to bring about blood clotting. Were it not for such mechanisms, we would be in danger of bleeding to death from every cut or abrasion. These disorders are not common and deserve only brief mention. Bleeding under the skin may result from weakening of the walls of small blood vessels, resulting in purple patches on the skin (*purpura*). Prothrombin deficiency due to deficiency of vitamin K may result in failure of the blood to clot. The remedy is to eat foods high in this vitamin (spinach, cabbage, egg yolk). *Hemophilia* occurs in certain families, passed on by the female but affecting only the male. Blood fails to clot due to deficiency of thromboplastin. The illness is rare and can be avoided by care in mating.

(c) *White cell disorders.* The function of the white cells is to defend the body against infection by devouring the bacteria. When infection occurs there is normally a great increase in the number of white cells in the blood (leukocytosis). This is a normal and proper reaction to infection. In *leukemia* this increase in production of white cells occurs without being needed by the body to combat infection. The symptoms of leukemia are at first those of anemia (fatigue, loss of energy). This is because the enormous numbers of white cells crowd out the red cells. Hemorrhages into the skin and mucous membranes (purpura) are common. The lymph glands may be enlarged and there may be severe bleeding and swelling of the gums. Microscopic examination of the blood shows that the white cells have increased enormously (from a normal value of up to 10,000 cells per milliliter to as many as 200,000 per milliliter). The course of this disease may be delayed by use of drugs or irradiation but at this time there is not a reliable cure.

9. Skin disorders

The skin, that sensitive envelope in which the rest of the body is wrapped, is one of the surest indicators of the general state of health. Skin suffers from two kinds of disorders, infections and allergic reactions. The commoner skin infections are as follows:

(a) *Impetigo.* This skin disease, frequently met with in infants and young children, is caused by infection with staphylococci or streptococci. It begins as a crop of small blisters filled with clear serous fluid (vesicles). They are commonest around the mouth and behind the ears, may spread to the scalp. The vesicles later form crusts of yellowish color, having a typical "stuck on" appearance. The infection is very contagious and every care must be taken to avoid spread of the disease. The thick crusts can be removed by bathing and use of starch poultices. Local application of mercury ointment or penicillin may help.

Related to impetigo are barber's rash (sycosis), an infection of the follicles of the beard region, and boils which are commonly found on the back of the neck. Both infections are signs of debilitation. Injections of penicillin usually clear the condition, but severe boils (carbuncles) may have to be lanced to drain the accumulated pus.

(b) *Acne.* This skin disease is common in adolescents, rarely seen in those over thirty. The symptoms are pimples and blackheads (comedos). In severe cases pustules may form which leave permanent scars. The best cure is washing the skin, avoiding greasy skin lotions and expressing blackheads *not* with the fingers (which may lead to worse infection) but with a special comedo extractor.

(c) *Scabies.* This very contagious skin disease is caused by the female itch mite which burrows into the skin, where it lays its eggs. The eruption is commonest on the wrists, webs of fingers and buttocks. Each eruption contains a mite which can sometimes be seen as a tiny white speck at the end of the burrow. Because of extreme itching a red papular eruption with scratch marks is present and secondary infection may result in impetigo. Treatment is by careful washing with hot water and application of a benzyl benzoate emulsion all over the body from the neck down.

(d) *Lice*. Lice come in three models, head louse, body louse and crab louse. The head louse lays its eggs (nits) in the hair. They can often be detected as small white specks. The body louse typically lives in the seams of the underclothes, the crab louse affects the pubic region, the anus and the armpits. Bites of lice cause severe irritation and scratching and may lead to impetigo. Application of DDT powder or emulsion will kill lice (it is one of the few permissible uses of this substance). Benzyl benzoate is also effective. Hot ironing of seams of clothes will kill body lice.

(e) *Fungi*. Skin infections due to fungi are known as ringworm. Ringworm can affect the hair, the skin or the nails. One of the commonest forms is athlete's foot, a fungus growth between the toes. The disease is infectious and measures should be taken to prevent its spread. Socks must be thoroughly boiled. The feet should be cleansed and dried and treated with one of the many powders or ointments available. Ringworm can also infect the groin, especially in hot weather. Ringworm of the scalp is usually seen in children. It causes a scaling bald patch, the hairs breaking off just above the level of the scalp. Ringworm of the nails is a resistant infection causing the base of the nail to become swollen and septic. It is often seen in women who frequently immerse their hands in water. Benzoic or salicylic ointment may help clear the infection.

(f) *Cold sore and shingles*. Both of these skin disorders are caused by viruses. The cold sore commonly appears on the lip, frequently after exposure to sunlight. Shingles appears as small blebs or vesicles which dry to form crusts. It may occur over one eye or, girdle fashion, around one side of the chest. Cold sores can be trusted to clear by themselves. Shingles, especially when the cornea of the eye becomes ulcerated, should be treated with aureomycin. There is often severe pain in shingles, which is really more a disease of the nerves than the skin. Get medical aid.

The commonest allergic diseases of the skin are:

(g) *Rhus dermatitis.* Parts of America are cursed with three members of the genus *Rhus:* poison oak, poison ivy and poison sumac. The poison is present in all parts of the plant but, unlike that of the stinging nettle, does not make its presence felt until about 2 days after exposure. Itching and redness followed by vesicles are the symptoms. The fluid in the vesicles is also poisoned. A particularly dangerous form of poisoning occurs when people burn this plant material and inhale the smoke. In such cases the lining of the lungs swells and great difficulty is experienced in breathing. The condition can be ameliorated with cortisone injections. In severe cases, get help. The best treatment is to avoid the plants. Wash parts of the body exposed to the plant as soon as possible after exposure.

(h) *Eczema.* The lesions of eczema are like those of Rhus dermatitis but the cause of the condition is not easily determined. Sometimes it is due to an irritant. Local treatment with calamine lotion or with zinc paste may help control the condition. *Urticaria* is a skin lesion resembling that produced by the stinging nettle, also called nettle rash. It is an allergic reaction which can be produced by various things: injection of serum, eating certain foods, taking certain drugs, the bites of some insects. It is important first to find the cause. Treatment with antihistamine drugs may help reduce symptoms.

10. Venereal diseases

"CLAP CLAP FOR PATTICLAPE—Patticake had her strange bladder pain diagnosed in town the other day. Yes, yes, we have genuine gonorrhea right here on the farm. It was a riot watching her trying to get up the gumption to tell George that she had given it to him the night before. Ha Ha."

This comment on life in the commune is from the *Last Whole Earth Catalog* (p. 181). It is not really so funny. The rate of increase of gonorrhea and syphilis among the "sexually liberated" who have spurned what they call bourgeois morality is attaining epidemic proportions. This epidemic could easily be controlled if a few simple precautions were taken. The fact remains that they generally are not, so venereal disease spreads widely. Two diseases of this type are prevalent in the temperate zones.

(a) *Gonorrhea.* The infectious agent is the gonococcus. Acute inflammation of the urethra (the tube leading from the urinary bladder to the exterior) is the chief symptom in the male. In the female the infection tends to locate in the neck of the uterus. Urination becomes painful. There is a thick, purulent yellow discharge. The infection may spread to the prostate in the male and to the uterine tubes in the female, often resulting in sterility. Fortunately, the gonococcus is generally susceptible to penicillin, though resistant strains are now appearing. Get help fast!

(b) *Syphilis.* The cause of the disease is a spirally twisted spirochete, *Treponema pallidum.* It is spread by sexual intercourse or by careless handling of infected material. The outstanding sign is the syphilitic sore or chancre, commonly occurring on the penis in the male and the labia of the female. It bleeds easily and is hard, which differentiates it from soft chancre. If left untreated the sore heals in a few weeks and the secondary stage begins with rashes, enlargement of lymph glands, mouth ulcers and general malaise. If still left untreated the disease proceeds to the third stage with deep ulcers, bone inflammation, damage to the heart and nervous system.

Early treatment of the disease is therefore vital. Any sore or chancre on the genitals should be examined by a physician. Care should be taken to prevent the spread of the disease by handling of infected material. The smallest scratch on the skin is sufficient to allow the spirochete to enter the body. All dressings should be burned and the patient's feeding and washing utensils kept separate. The spirochete is very susceptible to penicillin and massive doses of 10 million units or more may be given over a period of several weeks. Syphilis is one disease that modern medicine could have conquered were it not for the fools with which the physician has to deal. The folly, in this case, lies in failing to report the disease and spreading it to others. Syphilis is curable, folly is not.

11. Cancer

There are so many forms of cancer that it is probably incorrect to refer to it as a single disease. It seems highly probable that some forms of cancer are caused by a virus. Others result from injury, exposure to radiation or certain chemicals.

Cancer is a disease of cells. The cells involved multiply without regard for the needs of the body. Uncalled-for multiplication does not, by itself, result in cancer. It may produce a benign growth, wart or cyst, which poses no serious threat to the body. The growth is called cancerous or malignant when the cells involved break away from the original tumor and form growths in other parts of the body, particularly in the lymph glands. These secondary growths are called metastases. Once a cancer has metastasized it becomes difficult to remove. The successful treatment of cancer, therefore, depends on early diagnosis.

The danger signals of cancer are as follows: (1) Unusual bleeding or discharge. (2) A lump or thickening in the breast or elsewhere. (3) Any sore that does not heal. (4) Persistent change in bowel or bladder habits. (5) Persistent hoarseness or cough. (6) Persistent indigestion or difficulty in swallowing. (7) Change in a wart or mole. (8) Any unexplained change in body function, including a persistent feeling of ill health.

Two common forms of cancer in the female can be fairly easily detected. Cancer of the breast can generally be felt if the woman lies down and feels her breasts for lumps of hard masses. Cancer of the cervix (mouth of the uterus) can be detected by a "PAP smear," which reveals the presence of cancer cells. Skin cancer is reasonably easy to detect. It often occurs in areas of the skin exposed to sunlight, for which reason overexposure of the skin to sunlight is not advisable. A complete cancer check can be carried out by a physician, involving examination of blood, pelvic organs, lower bowel, stomach and even lungs, but not everyone is prepared to go to all the trouble and expense to be reassured on this matter and even the most conscientious physician cannot check all the organs of the body. Fear of cancer, in some people, becomes a disease in itself.

12. Mental illness

The term mental illness covers a number of disorders which involve the emotions rather than the intellect. These disharmonies are very poorly defined and may range from behavior which is classified as merely eccentric to obvious madness. It is impossible to draw a line between physical and mental illness. All mental illness affects the body in one way or another. The mind cannot exist in a vacuum.

(a) *Schizophrenia.* This is by far the commonest of the serious mental illnesses. The cause is unknown but an increasing mass of evidence suggests that the schizophrenic is a victim of a poison brewed in his own body. This poison, so goes the theory, works rather like LSD or mescaline. It distorts perceptions of the outer world. The schizophrenic lives in a world of illusions, frequently unpleasant and often terrifying. When the illusions are frightening, the illness takes the form of paranoid schizophrenia. The victim feels persecuted and endangered and sometimes attacks others under the influence of the delusions.

In catatonic schizophrenia the victim of the illness turns away from the outer world, becomes mute and motionless, entirely preoccupied with some inner fantasy. The motionless periods may alternate with episodes of violence.

Schizophrenia is potentially a dangerous illness and must be taken seriously. One school of thought, which has adopted the ponderous title of *orthomolecular psychiatry,* recommends heavy doses of vitamins, especially niacin and niacinamide, as a method of correcting the chemical errors from which the symptoms of schizophrenia result. Unfortunately, this brand of treatment seems to work only on certain people. Others are unaffected or even made worse. There are, however, a number of other anti-schizophrenic agents, the efficacy of which has been fairly well established. One thing that will *not* cure schizophrenia is psychoanalysis. Freud himself admitted this, yet people still spend fantastic amounts of time and money seeking a cure for the illness through these foggy procedures.

(b) *Paranoia.* This illness can be masked and for this reason can be particularly dangerous. The paranoid individual is sane outwardly and behaves quite normally except in one area. This area of the mind houses a complete system of delusions centering on the idea that someone is trying to damage or destroy the subject. No amount of reasoning or reassurance will shake these delusions. They may, however, be considerably blunted by suitable medication, provided the paranoid will admit he is ill and accept treatment. Unfortunately, this is one thing the paranoid generally refuses to do.

(c) *Psychopathic personality.* This is another condition that is extremely difficult to treat. Like the paranoiac the psychopathic deviate appears normal. He wears the mask of sanity. He even, apparently, can distinguish right from wrong. But he seems quite unable to use this knowledge, so is always in trouble. He leaves behind him a trail of wreckage, broken marriages, debts unpaid, crimes big and little. When hauled before the judge he will plead insanity. In a mental hospital he will act sane and work for his release. Once out he will re-enact all the same stupidities. He never learns and should, for the welfare of all concerned, be permanently removed from contact with society.

(d) *Manic-depressive psychosis.* The normal mood swing between elation and depression can, in some people, become exaggerated to the point that they become quite unbalanced. When elated these people are megalomaniacs, when depressed they are potential suicides. In the manic phase they may involve themselves in rash business deals or undertake financial obligations they cannot possibly fulfill. The illness creates problems both for the sufferer and for his friends and family. Medication may help to moderate the mood swings.

(e) *Depression.* Depression goes hand in hand with a number of physical symptoms. There is commonly sleep disturbance, loss of appetite, constipation, loss of sexual urge, a generalized inertia and loss of interest. In addition, there are feelings of guilt and of unworthiness. The danger in these states is always suicide. The condition can often be ameliorated by electroshock treatment or with certain antidepressive drugs. It almost certainly has a biochemical cause but what this cause is we do not yet know.

FIRST AID

The following accidents can be a threat to life and demand emergency treatment:

(a) *Blockage of airway.* The human body must have a constant supply of oxygen. If anything obstructs flow of air to the lungs death may follow in a few minutes. The airway consists of nose and mouth, throat, windpipe (trachea) and bronchial tubes. It may become blocked as a result of severe injury to the face or throat. The tongue in an unconscious person may fall back into the upper throat. Some foreign body may obstruct the airway. To clear it, place the patient on his back on a flat surface. Remove any foreign matter from the mouth or throat. Check the tongue. If it is blocking the throat, pull it forward.

In the case of a child who is choking because of a foreign body lodged in the airway, hold the child across your knees head down and administer a blow between the shoulders. If the child is breathing, do not use this procedure. Take him or her to the nearest medical facility.

(b) *Respiratory failure.* When breathing stops, unconsciousness and death will rapidly follow unless oxygen can be forced into the lungs and breathing started again. The most effective way of getting respiration restarted is the mouth-to-mouth method. The following steps should be taken: (1) Place the victim on his back, elevate shoulders with a pillow, rolled blanket, coat, etc. (2) Check the airway, remove obstructions. (3) Tilt back the head to open the airway. (4) Lift the lower jaw by grasping the jaw angles and lifting until the lower teeth are higher than the upper teeth. This pulls the base of the tongue away from the throat. (5) Take a deep breath and exhale into the victim's mouth, nose or both. Watch the chest to see if it rises. If it rises this means the lungs are filling. (6) Break the contact, listen for the exhalation. (7) Continue at the rate of 12 inflations per minute for adults, 15 for children, 20 for infants.

(c) *Drowning.* Drowning occurs when the lungs fill with liquid. Quite small amounts of liquid may suffice. A man can drown in his own body fluids. Death in drowning does not result solely from suffocation. The presence of large amounts of water in the lungs has serious effects on the heart and circulation and death in drowning may be from heart failure. Resuscitation will work in most cases if the victim has been rescued before the terminal gasp fills the lungs with fluid. Mouth-to-mouth breathing should always be attempted.

(d) *Bleeding.* There are various kinds of bleeding. A slight cut or graze causes blood to ooze from the capillaries. If the bleeding is from a vein, dark blood will flow at a steady rate under low pressure. Bleeding from an artery is swift, the blood is bright red and spurts from the blood vessel. Blood flowing in a steady stream or in large spurts indicates a serious condition and steps must be taken rapidly to control the hemorrhage. A large compression bandage over the wound may stop the flow. If this fails pressure may be applied to one of six points. The point selected must be between the heart and the bleeding area. The pressure points are:

(1) Temporal artery in the hollow joint just in front of the ear. (2) Facial artery, in the crevice about an inch from the angle of the jaw. Pressure is applied upward against the jawbone. (3) Carotid artery, located deep and back on each side of the windpipe. The artery must be compressed with the fingers against the bones of the neck. (4) Subclavian artery, located in the hollow near the connection of the collarbone and sternum. Pressure is applied by pushing the artery against the first rib. This will close off part of the circulation to the arm. (5) Brachial artery, on the inner side of the upper arm about 3 inches below the armpit. Pressure is applied against the bone of the upper arm. (6) Femoral artery, midway between the crotch and the point of the hip. Pressure can be applied with the heel of the hand against the bone of the pelvis.

A tourniquet should be applied only if the limb in question has been severed by the accident or so badly crushed to be beyond repair. Much harm can be done by amateurishly applied tourniquets. Use a strong, wide piece of cloth, tie a half knot, place a short stick in the half knot, tie a full knot over the stick. Twist the stick just tight enough to stop bleeding. Loosen the tourniquet for a few seconds every 15 minutes.

(e) *Strains, sprains and fractures.* A *strain* is the result of putting more tension on a muscle than it can take, so that the muscle or tendon is torn. The commonest type of strain affects the muscles of the lower back. It results from lifting a heavy weight with the back muscles instead of bending the legs and using the powerful leg muscles to help raise the weight. The signs of such back injury are severe spasms in the lower back, pain and often an inability to move. Treatment involves application of warmth to relax the spasm, and application of supportive strapping to the back.

The objective of strapping the back is to take the strain off the affected muscles, tendons and ligaments. Adhesive tape 3 inches wide should be used and the area covered should extend from the top of the pelvic bones to the lower edge of the ribs. The strapping is, of course, best applied by a doctor but in an emergency temporary strapping will help the victim to move about without further damaging the back muscles.

A *sprain* is an injury to the ligaments that hold a joint together. The commonest sprains are those of the ankle, knee and wrist. Most sprains are caused by a sudden twist or wrench, as when the foot is caught in an obstruction while running. Treatment of a sprain involves rest, immobilization and support. Cold applications for the first 24 hours will help counteract swelling. After 24 hours hot applications will help absorption of the swelling. A firm elastic bandage offers the safest means of support. Application of adhesive tape to immobilize a sprained joint is best left to a doctor. It is also desirable to have severe sprains x-rayed to determine whether there has been bone fracture.

Dislocations may occur in a ball-and-socket joint such as the shoulder when, due to a sudden strain, the ball simply slips out of the socket. The signs and symptoms are rigidity, loss of function, the unnatural shape of the affected joint, severe pain and marked swelling. The commonest dislocations are of the jaw, shoulder, elbow, hip and finger. It is best to immobilize the part with a splint if this is practical and get the victim to a doctor. Amateurish attempts to reduce the dislocation may make matters worse.

Fractures occur when a bone is put under so much strain that it breaks. In a simple fracture the ends of the broken bone do not penetrate the skin. If one or both ends of the bone puncture the skin so as to make an external wound the break is called a compound fracture. A green-stick fracture is one in which the bone is not broken through completely. Such fractures are commonest in children whose bones are still pliable. It is often very difficult to decide whether or not a bone has been broken. Acute tenderness at the site of the break and considerable pain and swelling will always occur. Deformity of the broken part and a grating sensation in the affected area, shortening of the affected limb, or a wound with bone projecting through it are sure signs of a fracture.

The following precautions should be taken when dealing with broken bones: (1) Never attempt to set a compound fracture. Simply cover the open wound with a thick sterile gauze pad and bandage it in place. (2) Try to avoid changing the position of the injured person until the exact nature of his injuries is known. (3) If you know where the fracture has occurred, try to apply splints to immobilize it before the victim makes any movement. (4) Never attempt to transport a victim in any but a reclining or semi-reclining position. Much harm can result from jamming an injured person into the back seat of a car and rushing him to a hospital instead of waiting for an ambulance, or at least for a station wagon. (5) In the case of suspected back or neck fracture, do not move the victim at all. Permanent paralysis can result.

First aid as applied to fractures must be designed to prevent further damage while the victim is transported to hospital or to a doctor. Admittedly, it is much better to bring the doctor to the victim but in out-of-the-way places this may not be possible. An excellent account of how to proceed in such cases is contained in *Emergency Medical Guide* by John Henderson, M.D. (McGraw-Hill, New York, 1963). Any true Whole-earther who proposes to live far from medical aid should own a copy of this book. It could save his or her life. Such adventurous Whole-earthers should also learn how to obtain emergency rescue help in the area where they propose to live.

(f) *Heat and cold injury.* The body can be damaged both by excess heat and by excess cold.

Burns are of three degrees: A first-degree burn involves intense reddening of the skin, as in mild sunburn or a moderate scald. It will not cause scarring and will heal of its own accord.

A second-degree burn will cause reddening of the skin followed by blister formation. Such burns do not cause scarring and do not require skin grafts. The skin should be gently cleansed and soaked in a solution made by adding 2 tablespoons of baking soda to a quart of boiled tap water which has been allowed to cool. After about 20 minutes of soaking, the skin can be dried and treated with a good antiseptic burn ointment. A second-degree burn will heal more quickly if blisters are drained. This can be done by painting the blister with non-irritating antiseptic and opening it near the base with a needle sterilized by heating over a flame.

Extensive second-degree burns may be accompanied by a certain degree of shock. Give fluids by mouth, keep the victim comfortably warm, cover the burned areas with sterile dressings and, if it seems necessary, remove the victim to a hospital.

Third-degree burns involve the entire thickness of the skin. In man such burns cannot repair themselves. A rabbit can repair a third-degree burn. A man can only pull the edges of the damaged area together, which results in scarring. Therefore third-degree burns must be treated with skin grafts. A person with such severe burns should be given fluids, sedated if possible, and taken as quickly as possible to a hospital.

A few don'ts should be remembered in connection with burns. Don't apply butter, tannic acid, picric acid or boric acid. All these burn treatments have been praised at one time or another. They mostly do more harm than good.

Frostbite is similar to burn injury because it involves local injury or death of tissue. Instead of being burned the tissue is frozen. Dead white skin and loss of all sensation is the surest sign of frostbite. Take immediate measures to restore circulation to the frozen part. Remove boots, socks, gloves, immerse the part in warm water (90–104° F) until thawed. Give as much hot, stimulating fluid (soup or coffee) as the victim will take. Encourage the victim to move the part as soon as possible but avoid rough handling. Avoid putting pressure on the frostbitten area. Keep it covered with a dry sterile dressing. *Do not rub the part with snow.* This is about the worst possible way to treat frostbite. The most serious consequence of frostbite is gangrene. This means that the frozen tissue has actually died and is starting to rot. If gangrene sets in, get medical help.

There are three other emergencies that may arise as a result of heat. *Heat cramps* are severe muscle cramps and pain accompanied by faintness and dizziness. They result from excessive sweating and loss of body salt and may be remedied by taking salt tablets.

Heat exhaustion is characterized by fatigue, lassitude and faintness followed by profuse clammy perspiration, whiteness of skin, loss of consciousness (brief), cold clammy skin, thready pulse, shallow breathing. Remove the victim to as cool a place as possible, cool the body with moist cloths, give salt tablets.

Heat stroke (sunstroke) is far more serious than heat exhaustion. In heat stroke the body's self-cooling mechanisms break down. A body temperature of over 105° F, a red, hot, dry skin, collapse or unconsciousness all occurring during a spell of very hot weather indicate heat stroke. The condition constitutes a grave emergency. Body temperature must be lowered quickly or irreparable damage can be done to brain, kidneys or liver. Place the victim in a tub of cold water with cracked ice if possible. If this is not available, cover the naked body with wet sheets and use a fan to accelerate cooling. Administer rectal irrigation of ice water. Get medical aid.

(g) *Poisoning.* Accidental poisoning is best prevented by keeping poisonous substances out of reach of children. Poisons are not only taken up through the stomach. They may be inhaled as vapors (insecticides, cleaning fluids, gas from leaky appliances, automobile exhaust). They can be absorbed through the skin. They can be injected by insects, spiders, snakes, etc. The main types of poisoning are as follows:

(1) *Corrosive acids.* Examples are sulfuric acid (car batteries), hydrochloric acid (metal cleaners), nitric acid (industrial cleaning solutions), oxalic acid (cleaning solutions), carbolic acid (disinfectants). For all except carbolic acid, give lime water, chalk or 1 teaspoon of baking soda in a glass of water to neutralize acid. For carbolic acid poisoning, give alcohol in the form of whisky or wine. If there is no alcohol, give white of egg beaten up. Induce vomiting with soapy water or syrup of ipecac, a good substance to have on hand in families with children. Get medical aid.

(2) *Corrosive alkalis.* Examples are potash (lye), caustic soda (soap making), quick lime, ammonia. Give large amounts of lemon or orange juice or equal parts of vinegar and water to neutralize alkali. Give milk or raw eggs or food oil to soothe and protect the tissues. Do not induce vomiting. Get medical aid.

(3) *Petroleum distillates.* Examples are kerosene, gasoline, turpentine, paint thinner. Give 1 to 2 ounces olive or other vegetable oil. Do not induce vomiting. Get medical aid.

(4) *Depressant drugs.* These are barbiturates and tranquilizers. The symptoms are drowsiness, slurred speech, shallow breathing. Induce vomiting if you can't get a doctor with a stomach pump. Give large amounts of warm water to wash out the stomach, administer artificial respiration if breathing is failing, keep the victim warm, remove to hospital.

Opium derivatives, heroin and morphine, produce similar symptoms but the pupils of the eye are contracted to pinpoints. Treatment is the same as for barbiturates.

(5) *Irritants.* Examples are arsenic (rat poisons, weed killers), copper (plant sprays, rat poisons), iodine (antiseptic), lead (paint, putty), mercury (plant sprays), phosphorus (matches, rat poison), silver nitrate (inks, solutions). For treatment, give an emetic, wash out the stomach, give 2 ounces Epsom salts in 8 ounces water. For iodine poisoning, give a mixture of starch and water. For silver nitrate, give a solution of ordinary salt.

(6) *Gases.* Carbon monoxide is given off in car exhaust and may be generated by a burning stove with insufficient air. Symptoms are headache, yawning, dizziness, pounding of heart, lethargy, stupor and a bright cherry-red skin color. Move the victim into fresh air, start artificial respiration if breathing is weak, call a first-aid squad for oxygen.

Carbon tetrachloride is present in cleaning fluids. The vapor is very harmful. Symptoms are headache, dizziness, nausea, vomiting. For treatment, get the victim into fresh air, give oxygen or artificial respiration, hospitalize.

Cyanide is one of the most rapidly acting of poisons. The gas may be given off from some preparations used for killing moles or gophers. The victim collapses and death follows from respiratory paralysis. All cyanide preparations should be handled with extreme care.

(7) *Chemicals in the eye.* Wash the eye with clear water, with eyelid held open and water flowing away from the unaffected eye, for 5 minutes. Get medical help.

(8) *Poisonous bites and stings.* Stings from wasps, hornets, yellow jackets and bees become serious only in people who have been sensitized to the venom. Such people may develop *anaphylactic shock*. Symptoms are severe difficulty in breathing, blueness of skin, cough, headache, unconsciousness. Or there may be nausea, vomiting and diarrhea. If medical aid is not immediately available the drugs isoproterenol or epinephrine can be used to tide the victim through the emergency. Kits can be obtained containing these materials and sensitive individuals should possess one in case help is not available. People not sensitive should scrape off the bee sting (which always remains in the skin) and treat the area with a strong solution of Epsom salts.

Black widow spiders are about ¾ inch across with a red hourglass-shaped marking on the underside. A bite produces severe abdominal pain due to muscle spasm, partial collapse, generalized swelling of face and extremities. For treatment, make a crisscross incision over the bite and suck out the poison. Apply a tourniquet just above the bite but not so tight as to cut off arterial blood. Get medical aid if symptoms become alarming. An antivenom is available and calcium lactate injections will help relax the muscle spasms.

Snakebite in the United States is generally inflicted by pit vipers (rattlesnake, moccasin, copperhead). The two fangs of these snakes leave distinctive puncture wounds. There is immediate severe pain, swelling and dark purplish discoloration of the skin, weakness, shortness of breath, dimness of vision, rapid pulse, nausea and vomiting. Bites by a coral snake (rare) result in great drowsiness and then unconsciousness. The victim should be kept as quiet as possible to slow down absorption of the toxin. A crisscross cut should be made over the wound and the poison sucked out. If you have a snakebite kit, use the suction cup. To slow spread of poison, put a constricting band about the extremity just above the bite and just tight enough to stop flow of blood through the veins. Get medical aid.

(9) *Poisonous plants.* By far the most dangerous of poisonous plants is the death cap, *Amanita phalloides*. This fungus is picked because, in the button stage, it looks like the common mushroom. This poisoning is generally fatal; therefore, never eat any white fungus unless the gills are pink or chocolate color. Other poisonous plants are oleander, aconite, Jimson weed (thorn apple). In all cases of poisoning by plants the first thing to do is to try to get the plant material out of the stomach with an emetic (such as mustard and water). Once the poison has been absorbed it is necessary to summon medical aid as antidotes for such poisons are not likely to be available except to a doctor.

In order to be prepared for accidental poisoning, especially in the case of children, find out the telephone number of the Poison Control Center nearest you. A local hospital or the telephone book will supply this information.

CHILDBIRTH

A strong "back to nature" movement has developed around the process of birth, particularly among Whole-earthers. The argument goes more or less as follows: Childbirth is a natural process. It is not some sort of disease and it should not have to happen in a hospital, amidst all the paraphernalia of disease and death. The proper attitude toward childbirth should be one of joyful expectancy, not one of tense apprehension. Peasant women give birth to their babies in the morning and are back at work in the afternoon. So let's have the baby at home and make it a festive occasion and have done with all this nonsense about anesthetics and obstetrics.

All of which is very cheerful and optimistic but fails to take into account one unfortunate fact of evolution: that, of all the mammals, the human female is the most badly designed for reproduction. Her upright posture puts her in danger of developing varicose veins every time she gets pregnant. Her pelvic opening is only just big enough to allow the child to be expelled. The idea that "nature knows best" does not always hold when it comes to childbirth. Nature bungles with sufficient frequency to make it highly desirable that a woman in labor should be within reach of medical aid. A correspondent to the *Last Whole Earth Catalog*, who had herself experienced a difficult, dangerous childbirth, puts it this way: "I dislike many, many of the fruits of modern civilization. But modern obstetrics is one I would sooner not dispense with."

So, without in any way reducing her enthusiasm for natural childbirth, let the modern female Whole-earther realize that the process of giving birth does present certain hazards and that expert help is not to be lightly rejected. However, if it does happen that, for some reason, help cannot be obtained or something prevents the mother from getting to a hospital, then the following steps can be taken to ease the process of birth:

(1) The mother should empty the rectum with the aid of an enema. Contamination of the birth area by involuntary expulsion of feces does not help matters.

CHILDBIRTH 243

(2) Place a rubber sheet on the bed and cover with a clean sheet. The mother should wash her genital area with soap and hot water, put on a clean nightgown and heavy socks to keep her legs warm.

(3) The first stage of labor brings the baby out of the womb into the birth canal (fig. 34). The second stage brings it out of the birth canal into the world. The vulva opens and the top of the child's head can be seen. Usually the baby's head will be pointing downward. The person assisting the birth can ease delivery of the head (the most critical part of the birth process) by cover-

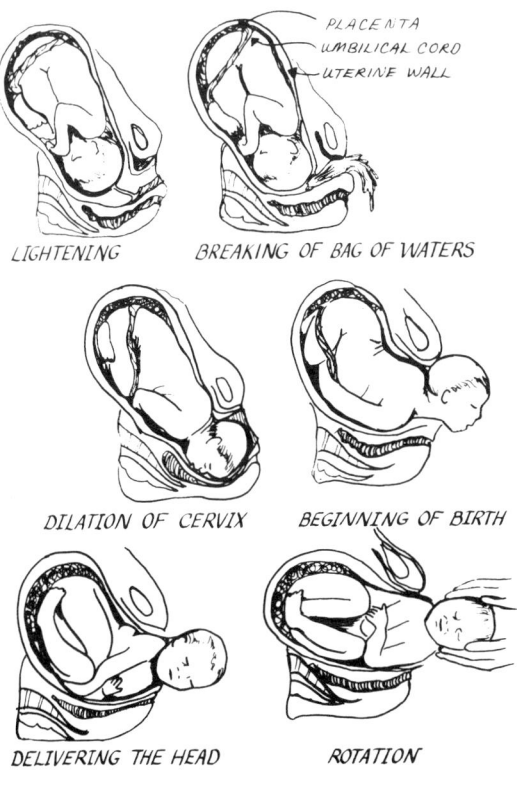

CHILDBIRTH

FIGURE 34

ing his right hand with a freshly laundered towel and pressing in on the mother's skin between the anus and the opening of the vulva. The assistant can feel the baby's chin, gently ease it up and out through the opening.

Once the head is born the rest of the birth is easy. The baby will slip out, bluish in color and covered with slime, still attached to the mother by the umbilical cord. Gently grasp the baby by the ankles and hold it upside down. Because it is very slippery, wrap the ankles in a towel or hold one ankle between thumb and index finger, second between index and remaining fingers. The assistant should use his other hand to wipe the baby's mouth and nose with a sterile pad and then pat him between the shoulder blades. At this point the baby should start to breathe, uttering its first cry.

The baby should be laid across the mother's abdomen, its head a little lower than the rest of its body. To tie off the umbilical cord, take a piece of sterile cloth tape and tie a knot 4 or 5 inches from the point where the cord joins the baby. Tie a second piece of tape 2 inches nearer the mother. Take sterilized scissors and cut the cord between the ties.

The baby is now a separate individual. It should be wrapped in a blanket and placed in its basket. (In hospital deliveries a silver nitrate solution is dropped into the eyes of newborn infants. This is required by law in most states to prevent infection from the gonorrhea germ, a cause of blindness. Check with a doctor about this.) Meanwhile the mother will be entering the third stage of labor, which generally occurs 10 to 20 minutes after the birth. This stage involves expulsion of the *placenta,* the special tissue through which the baby has drawn its food and oxygen. The assistant should not pull on the cord to hasten expulsion as this may rupture a blood vessel. Catch the placenta in a basin and leave it for the doctor to examine. If part of the placenta has remained in the womb it may cause trouble later on.

It is important to minimize bleeding after delivery of the placenta by massaging the womb to help it contract. The womb can be felt just below the navel as a large, firm mass about the size of a grapefruit. It should be gently grasped through the abdominal wall and kneaded between thumb and forefinger to help it contract. This contraction prevents excessive bleeding. Massage the womb until it is firmly contracted and danger of hemorrhage is past. Meanwhile, cover the vulva with two sanitary napkins, covered in a sterile towel. Keep the mother warm and comfortable and offer her a cup of tea. As soon as she begins to breast-feed the baby the contractions of the uterus will accelerate and that hugely distended organ will begin to go back to its usual size.

DEATH

The indications of death should be familiar to all those intending to live in remote places. Legally, only a coroner or a physician can certify that a person is dead. Practically, it may be necessary for one who is neither coroner nor doctor to know the symptoms of death. Extreme caution is necessary. People have been declared dead by bystanders and others only to be revived later by a rescue squad.

Onset of death may take three forms: coma (death beginning in the brain), syncope (death beginning in the heart), asphyxia (death beginning in the lungs). The signs of death are the absence of pulse, no visible breathing, no reaction to light by the eye pupils. Even these signs are not completely reliable. In addition, the following signs should be looked for: (1) *Algor mortis,* the cooling of the body which follows death. The degree of cooling indicates when death occurred and depends on the size of the victim and the temperature of the surroundings. (2) *Rigor mortis,* a tightening of the muscles following death. This first affects the facial muscles (within 3 to 5 hours) and gradually extends downward. Within 8 to 12 hours all the muscles are stiff. These signs are later followed by *cadaveric lividity,* a bluish or purplish discoloration on parts of the body. Finally, putrefaction occurs, the decomposition of the body.

Such are the outward signs of death. Of far more importance

to the true Whole-earther are certain questions relating to one's own death. A sickly, spiritually impoverished culture takes for granted that one lives as long as possible and a misguided medical science uses all its tricks to prolong the life of the body, regardless of whether the occupant of that body wants life prolonged. The true Whole-earther, if he wishes to die rightly, must understand what death involves and learn, so to speak, to live with his own death. The physical body is a vehicle in which something travels. When the body wears out, as it inevitably does, that something leaves it. To try to prolong the use of the vehicle beyond its proper span is a sign of weakness and lack of real understanding. Dying is part of the natural cycle, as natural as the fall of leaves in autumn.

The true Whole-earther, for this reason, should take steps to prevent his physical vehicle from falling into the hands of the misguided medics, whose efforts to patch it up and keep it running will merely cause trouble and expense. It is advisable, for this reason, to take steps well in advance to ensure that the misguided medics leave one to die in peace. The best way to do this is to follow the example of members of certain Indian tribes, to know the approach of one's death and go off to some place one has selected as suitable and simply die there. It is not especially difficult to die thus. All one need do is stop eating. One can leave a note stating where the body may be found.

There is, however, always the possibility that the vehicle may be disabled, perhaps paralyzed by a stroke, and that it will fall into the hands of the meddlesome medics and fiddled with and fussed over despite the objections of its owner. To guard against this eventuality the true Whole-earther may be well advised to prepare and sign the following statement (suggested in *Euthanasia and the Right to Death* by A. B. Downing, Nash Publishing Co., Los Angeles, Calif., 1970):

"I DECLARE AS FOLLOWS:
"If I should at any time suffer from a serious physical illness or impairment, thought in my case to be incurable and expected

to cause me severe distress or render me incapable of rational existence, then, unless I revoke this declaration or express a wish contrary to its terms,

"I REQUEST the administration of whatever quantity of drugs may be required to prevent my feeling pain or distress and, if my suffering cannot be otherwise relieved, to be kept continuously unconscious at a level where dreaming does not take place, AND I DECLINE to receive any treatment or sustenance designed to prolong my life.

"I ASK sympathetically disposed doctors to acknowledge the right of a patient to request certain kinds of treatment and to decline others, and I assure them that if in any situation they think it better for me to die than to survive, I am content to endorse their judgment in advance and in full confidence that they will be acting in my interests to spare me from suffering and ignominy, and also to save my family and friends from anguish I would not want them to endure on my behalf."

Even this may not suffice to restrain the meddling of the medics who take some weird professional pride in what they call "cheating death," as if death can ever be cheated. But it is always worth trying. A form similar to the one given, known as a "living will," is available from the Euthanasia Educational Council, 250 W. 57th St., New York, N.Y. 10019. Another valuable source of information for someone who wants a dignified and simple end is *A Manual of Simple Burial,* available from Celo Press, Burnsville, N.C. 28714, $1.50.

IV ENERGY

Man is the only animal on the planet earth who uses energy other than that generated by his own muscles. First, man learned to domesticate or, more accurately, to enslave certain animals which were stronger than he. Later he learned to utilize wind power and water power. Still later he learned to use steam power and finally to burn oil and its products and convert the energy liberated into mechanical power. From this there resulted the runaway technology that now pollutes the biosphere and threatens, by its own greedy excesses, to bring itself to a screaming halt in the fairly near future.

All of which brings us to the main point of this chapter, which is that the true Whole-earther must immediately attempt to find alternate sources of energy unless he is willing to return to the pre-industrial age and do without a number of handy gadgets. In this chapter I propose to study as practically as possible the main energy sources available to a group of people who wish to rely on renewable non-polluting sources of energy.

Before we consider these energy sources the true Whole-earther should know what power is and how it is measured. Power is amount of energy per second. Energy is the physical ability to do work. Work is done when a body is moved by a force. The relationships of various units of mechanical energy and heat are shown in table IV.

TABLE IV
Relations between units of mechanical work and heat

1 erg = work done when 1.02 milligrams is lifted 1 centimeter vertically against gravity at sea level
10 million ergs = 1 joule
1 joule per second = 1 watt
1,000 watts = 1 kilowatt
watts = volts × amperes
746 watts = 1 horsepower
1 kilowatt = 1.34 horsepower
1 kilowatt hour = power generated by 1 kilowatt flowing for an hour = 860 kilocalories
1 British Thermal Unit = 0.252 kilocalorie = 778 foot-pounds

It is necessary, in thinking about energy, to distinguish between heat and mechanical power. Heat can be converted into power through the agency of a heat engine. The efficiency of heat engines is limited by the Carnot cycle. This means that the amount of heat a heat engine can convert into mechanical work is proportional to the difference between the highest and the lowest temperature within the engine. So although electrical or mechanical energy can be turned into heat with 100% efficiency, heat cannot be turned into mechanical energy with an efficiency much greater than 40%.

A practical understanding of the energy crisis which may wreck our power-hungry culture within the next few decades depends on a clear realization of how much of one form of energy can be exchanged for another. Conversion of electric energy to heat is always costly. By the time all the losses are taken into account little more than 20% of the heat value of the original fuel can be recovered from the electricity which it is used to generate. Electric heating devices are, for this reason, to be avoided. If heat is required, use solar heat whenever possible. Wood, locally gathered, is the next choice. Methane generated from waste is another good source. If forced to use fossil fuels, use coal rather than oil or natural gas.

Electricity can be generated by solar power, wind power, water power or, via a heat machine, from methane or other locally obtainable fuels. But the *amount* of electricity which can be generated by these means is not large and must be reserved for running machines that cannot be operated in other ways.

We shall discuss four forms of energy: solar power, wind power, water power, power from organic wastes (principally methane). These sources of power are non-polluting and do not draw on irreplaceable fossil fuels. A sane civilization with enlightened leadership would concentrate all its know-how on developing these energy sources, storing the remaining supplies of fossil fuel as valuable chemical materials. But our civilization is not sane and certainly does not choose enlightened leaders. So the amount of money spent on research into solar and other forms of clean energy is negligible. Instead the government pours money into the study of the breeder reactor, a device using as fuel one of the most poisonous substances known and generating radioactive wastes which will have to be guarded for thousands of years to come.

From an objective point of view any civilization that puts its faith in breeder reactors would have to be condemned as collectively insane. One explosion, one act of sabotage, one accident would suffice to render large areas of the countryside permanently uninhabitable. Plutonium has a half life of 24,000 years and is so poisonous that little more than a microgram (one-millionth of a gram) will cause cancer in animals. If the present program is put into operation, as many as 500 shipments of the hellish stuff per week will be traveling around the country subject to accidents, hijackings or the attentions of various forms of homicidal lunatics. The government of the United States proposes to spend as much as $3.9 billion on the development of breeder reactors that use this deadly substance. This same government virtually ignores the clean, harmless solar energy that pours down onto the deserts of the southwest and could be harnessed by techniques already developed.

So it goes!

SOLAR POWER

1. Solar farms

The total amount of sun power that hits the earth's surface in any 24-hour period is staggering. Even a power-squandering society like that of the United States could comfortably run all of its machines on a minute fraction of this solar energy. It would need, according to Aden and Marjory Meinel of the University of Arizona, only 1 square mile of "solar farm" to supply the power needs of a city of 60,000 people. Five thousand square miles of collectors (a square with about 70 miles to a side) would supply the electric power needs of the United States in the year 2000. The land would not be rendered useless. By raising the sun traps on stilts cows or sheep could graze beneath.

The solar energy trapping techniques developed by the Meinels might be adapted to run small (1–5 kilowatt) electric generators by any Whole-earther willing to spend time on devising the necessary hardware. The sun trap itself is a sophisticated device. Solar radiation is focused by means of a Fresnel lens on a window area in an evacuated glass pipe. The remaining portion of the interior of the glass pipe is silvered to produce high reflectivity. Sunlight enters through the window and impinges on a steel pipe with a thin selective coating which absorbs the infrared portion of the solar rays. Very thin coats of certain combinations of materials such as silicon and silver will do this and the steel pipes so coated can be raised to a temperature of 1,000° F (540° C) by this means.

The heat so generated is conducted by a gas (nitrogen) to containers of a suitable salt which store the heat. This heat is used to run either a steam turbine or a gas turbine.

The biggest drawback to the use of solar energy for power

generation is its intermittent nature. The sun shines only when the sky is clear and the intensity of sunlight varies with the seasons. Winter sunlight, striking the earth at a low angle, is so diluted that it carries little energy. The further north one goes, the smaller the energy becomes.

Modern technology, however, offers a solution to this problem. During the summer months power can be generated that is far in excess of immediate needs. This surplus power can be used to create a fuel that can be burned during the winter months to generate steam and drive a conventional turbine. Such a fuel could take the form of hydrogen, generated from water by electrolysis. But hydrogen, unless liquefied, requires enormous containers for its storage, for which reason aluminum might prove a better fuel. Metallic aluminum in finely powdered form burns to give aluminum oxide. The oxide can be converted back into metal by electrolysis.

These studies by the Meinels offer to any Whole-earther with sufficient ingenuity an opportunity to experiment with solar generators. The technique is sophisticated but not entirely beyond the reach of the amateur. Essentially, it is a modernized version of the solar engines built as long ago as 1884 which used paraboloid reflectors to focus sunlight on a cylindrical boiler. Designing the solar heat collector is not really a difficult problem. The element that is lacking is the turbine which, for small installations, must be capable of operating at relatively low temperatures. Such a turbine could hardly use steam as its operating medium but might work fairly efficiently on air or Freon. The ideal machine would be the closed-cycle gas turbine, as manufactured by the Escher-Wyss Engineering Works in Zurich.

2. Photovoltaic cells

Methods of converting solar energy to mechanical work which involve heating a working medium, activating a heat engine and from its rotation generating electricity are obviously clumsy. Sunlight is electromagnetic radiation and every photon (particle of light) carries energy which is proportional to its frequency of vibration. Could we not, therefore, convert solar energy to electrical energy without going through this elaborate heat trap–heat engine–generator cycle? The answer is yes, but the technology involved is so complex that it is beyond the reach of the ordinary citizen. We are dealing with events at a subatomic level, with a flow of electrons through a carefully constructed atomic lattice. This lattice is made from crystallized silicon to which measured amounts of impurities have been added. The devices certainly work. They have no moving parts. They have about an 11% efficiency and they do not pollute the atmosphere. At the moment photovoltaic devices are about a hundred times too expensive to be used for generation of electricity on the surface of the planet earth. They are useful mainly for spacecraft.

3. Solar heaters

The most obvious use the Whole-earther can make of solar power is to heat his water and his house. This does not involve any advanced technology. The simplest kind of solar water heater involves only three elements: a storage tank for cold water, a heat trap, a storage tank for hot water. All these units can be easily assembled by anyone who is reasonably handy. Construction involves the following steps:

(1) Cold water tank. If your water heater is to work off a gravity feed the cold water storage tank must be higher than the heater. A satisfactory spacing is that shown in fig. 35. The tank can be filled via a ball valve from the main water supply. A 50-gallon oil drum will hold enough water.

(2) Absorber. This is the unit which absorbs the sun's rays. How large you make it depends on how much hot water you want. You can make it from copper tubing, galvanized iron pipe, black polyvinyl chloride pipe, or black polyethylene tubing. The important thing to remember is that hot water is lighter than cold so the cold water should enter at the bottom.

Fix the water pipes of the absorber in a wooden frame about 36 × 96 inches. Under the frame place a 3 × 8-foot sheet of galvanized steel (24 gauge) painted black to absorb radiation. Below the steel sheet, fix a sheet of insulating material such as fiberboard. To the top of the frame, fix a sheet of glass or sun-

254 ENERGY

resistant vinyl plastic. The frame is, in effect, a miniature greenhouse and concentrates the heat on the water pipes. In extra cold climates a double layer of glass may be necessary. Holes should be drilled in the bottom of the frame to allow condensed water to escape. Set the absorber on the roof facing south at an angle of 15° + latitude of your house (for example, at latitude 40° the absorber would be set 55° to horizontal).

(3) Hot water tank. Lead the water from the top of the ab-

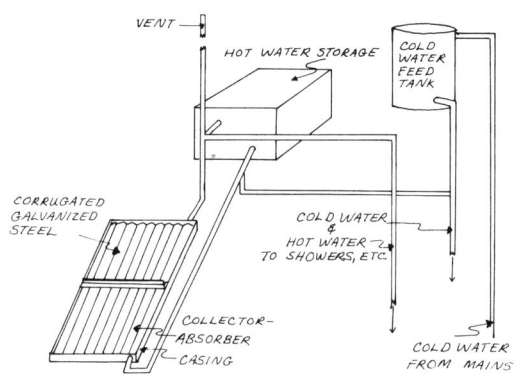

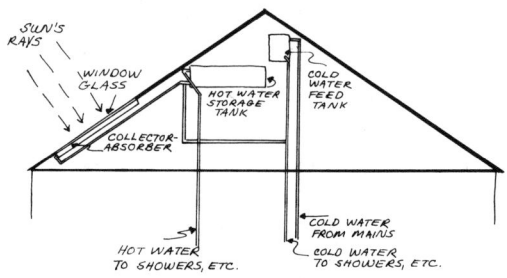

SOLAR WATER HEATER

FIGURE 35

sorber to the hot water tank, which should not be more than a foot above the top of the absorber. The tank should be insulated with fiber glass. Hot water from the hot water tank flows by gravity to all points, kitchen, bathroom, etc. If your house has an electric or gas water heater, don't throw it out. Solar heat is not entirely reliable and the alternate system will be necessary from time to time. You must, however, install a valve to cut out the solar heater before you switch to the alternate system if your solar water heater is designed to operate by gravity. Of course, it is perfectly possible to design a solar water heater that works directly off the mains under pressure. It is necessary in this case to make sure that all connections are strong enough to withstand pressure (40 pounds per square inch in most cases).

A solar house heater resembles a solar water heater. It consists of three elements: an absorber, a storage unit and a circulating unit. However, the absorber must be much larger than that required for water heating. The Whole-earther planning to construct a solar house can get ideas regarding dimensions from either the Dover House (designed by Dr. Maria Telkes) or from the Thomason House (designed and built by Harry E. Thomason in Washington, D.C.). The characteristics of these two houses have been compared (*The Mother Earth News*, no. 9, 1971). The chief features of each are shown in table V.

The following points should be borne in mind by anyone planning a solar heated house. (1) Try to build on a south slope. (2) Insulate the house carefully. (3) Excavate space beneath the house for heat storage. (4) Put in large south-facing windows and insulate them at night with shutters. Probably the best general design for the solar heated house is the one shown in fig. 19. Although water is a fine circulating medium it has one drawback: it freezes. Air has only a fraction of the heat storage capacity of water but it can be moved cheaply either through the heat storage bins or through the house. All one needs is a fan. Glauber's salt (sodium sulfate decahydrate, $Na_2SO_4 \cdot 10H_2O$) is capable of storing 8.5 times more heat than the same volume of water over a temperature range of 77 to 98° F. As the temper-

TABLE V
Characteristics of two solar houses

Dover	Thomason
Size of house	
Five rooms and bath	Six rooms
Cost of solar heating system	
$1,855 (in 1949)	$2,000 (in 1959)
Collector	
Area 720 square feet; blackened metal sheets behind double plates of glass	Area 840 square feet; blackened sheets of corrugated aluminum under one layer of polyester film and one layer of window glass
Heat transport	
Air circulating in space behind metal sheets	Water flowing over metal
Heat storage	
470 cubic feet; Glauber's salt sealed in 5-gallon cans stacked in columns to allow air circulation	1,600-gallon water tank in 50 tons of rocks in bin $10 \times 25 \times 7$ feet
Distribution of heat	
Warm air circulation	Warm air circulation

ature falls the salt crystallizes and emits heat in the process. The angle of the solar heat collector should be latitude $+ 15°$.

4. Solar stills

The use of sun power for distillation of water is an application that any Whole-earther might consider if he likes the idea of living in a desert by the ocean. There are miles of such desert coasts and they provide a healthful environment with plenty of opportunity for aquaculture and contemplation. All that is lacking is fresh water.

Solar stills can range in size from plastic spheres designed to supply shipwrecked mariners or downed airmen to installations covering several acres to supply whole communities. The principle of operation is in all cases the same. A thin layer of sea water is introduced into a shallow pan, the bottom of which is painted black and insulated on the lower side. Above the pan is placed a sheet of glass or plastic at an angle facing the sun. The whole structure constitutes a miniature greenhouse. Trapped solar heat raises the temperature of the sea water. Water vapor condenses on the cooler surface of the glass, trickles down and is collected in a trough at the lower edge of the glass plate. The trough communicates with a collecting vessel.

Solar stills in hot climates will produce about 1 gallon of water per 12 square feet per day. This means that rather large areas must be covered if water is to be used for irrigation as well as drinking. A cheap plastic-covered solar still is described in *How to Make a Solar Still, Plastic Covered* (Leaflet no. 1, Brace Research Institute, January 1965). Essentially, this still consists of a 120-foot length of stainless steel wire, 1,000 pounds breaking strength, supported at either end on a brick or concrete wall (fig. 36). The stainless steel wire can be sheathed in polyethylene tubing and is anchored at each end to metal stakes driven into the ground. A concrete foundation runs along either side of the still, to which is cemented a row of bricks. To hold the sea water the ground between the foundations is carefully leveled and covered with sand. This sand is then covered with black polyethylene liner 0.010 inch thick, 10 × 100 feet. The polyethylene covers a gutter made of sheet steel or sheet aluminum which runs inside the bricks along the whole length of the still. The still should preferably slope gently in the direction of the earth's equator and the sand should be so arranged that the water flows into a series of shallow dams. Depth of the water should be about 1 inch.

For the tent-like top of the still the Brace Institute recommends polyvinyl chloride film, ultraviolet-stabilized (from Gulf Oil of Canada, Montreal, Quebec), rendered wettable by treatment with a wetting agent such as Sun Clear (Solar Sunstill Inc., Setauket, N.Y. 11733). Sheets of glass will have a longer life than the plastic but are heavy to transport and fragile. The plastic tent must be held in place by a row of bricks to which a 1¼-inch

258 ENERGY

semicircular cement lip has been cast. This lip ensures that the drops of water on the inside of the plastic fall directly into the gutter. The still should be fenced to keep animals out and preferably protected from wind with a low windbreak. If sea water is used the concentrated brine should be flushed out daily to prevent salt deposition in the tray. The inflow should be adjusted so that 2 to 3 gallons an hour flow through this 100-foot-long still. A small amount of acid added to the sea water (sufficient to keep the pH around 6) will prevent scale formation.

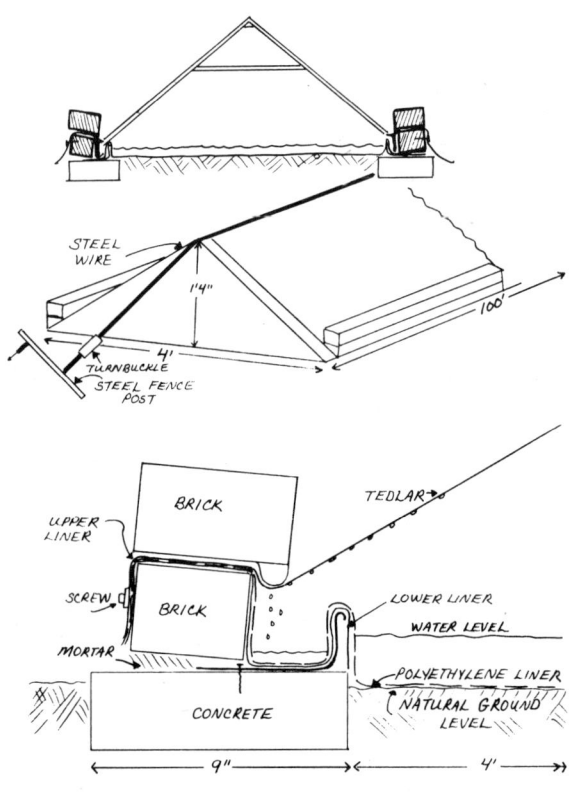

SOLAR STILL

FIGURE 36

A large glass and concrete solar still was constructed in Haiti in 1969 based on designs worked out by the Brace Research Institute (T. A. Lawand and R. Alward, *Plans for a Glass and Concrete Solar Still,* Technical Report #T.58). This still, which used the Savonius rotor wind pump to maintain the flow of sea water, was planned to provide an average supply of 200 gallons per day. The basic unit was simply a long plastic-lined trough 26 inches wide, above which were placed sheets of glass ⅛ × 30 × 18 inches. Fifteen such troughs 75 feet 7 inches long were arranged parallel to each other and all the fresh water was collected and stored in a single container. The concentrated brine was restored to the ocean. The glass plates were so arranged that they collected rain water as well as distillate from sea water.

Any Whole-earther who dreams of making the desert bloom by using water distilled from the ocean should study the methods worked out by the University of Arizona's Environmental Research Laboratory. Both at Puerto Penasco and at Abu Dhabi installations have been set up for growing vegetables in a desert region. No still on earth could supply enough water to compensate for the tremendous losses which a plant suffers in the desert. The solution is to prevent the losses by enclosing the plants in greenhouses made of plastic and inflated by small air pumps. Air in the greenhouse is kept 100% moisture-saturated by being circulated through a spray of sea water. Crops are planted directly in the desert sand and needed nutrients are added. Once enough fresh water has been accumulated to wet the sand, very little more need be added. The high humidity in the greenhouse prevents water loss. Aquacoms on the ocean could use such methods to grow such food as they are unable to get from the sea.

WIND POWER

Wind power is derived from the sun, an inexhaustible energy source though not a reliable one. The true Whole-earther who wishes to use the wind as a source of energy must first find out how much wind is available and from what direction it blows. We here on the western edge of California with an unobstructed view almost to the coast can rely on regular spells of windy weather with northwesterlies in summer and southerly winds in winter. Mornings will frequently be calm in summer but wind will blow in the afternoon as cool air from the ocean replaces air heated during the day.

The efficiency of a wind machine depends on certain aerodynamic principles which are fairly well known. No windmill can extract more than about 60% of the wind power that passes through it. Most fall far short of this value. There are two kinds of wind machine. The Savonius rotor, developed by S. J. Savonius, works on the same principle as a cup anemometer, has an efficiency of about 30% and is easy to build. The propeller- or sail-type wind machine is somewhat more complex to construct.

1. Savonius rotor

A simple wind machine for pumping water is described by A. Bodek of the Brace Experiment Station, St. James, Barbados, West Indies (Do-It-Yourself Leaflet no. 5, February 1965; for copies, write Brace Research Institute, Macdonald College of McGill University, Ste. Anne de Bellevue 800, Quebec, Canada). This machine can be easily constructed by anyone having access to welding equipment. It consists of three parts: rotor, drive and pump. Building it involves essentially the following steps:

(1) Procure two oil drums (45- or 50-gallon). Cut them lengthwise and weld them to form two troughs (fig. 37). Attach the two troughs spaced as shown to two discs cut from ½-inch plywood with ⅜-inch bolts. Pass a shaft (water pipe of 1¼-inch inside diameter) through the rotor, extending 6 inches beyond the end plates. Secure shaft to end plates with two flanged collars bolted to the plywood with four bolts and to the shaft with a ¾-inch bolt inserted in a hole drilled for this purpose. Support the rotor shaft in the frame on two self-aligning ball bearings. Attach two adaptors, one to each end of the shaft, to attach the bearings. Make the lower adaptor long enough to pass through the frame and the eccentric on its lower end. Balance the rotor carefully before mounting in the frame by placing it horizontally between

WIND POWER 261

two straight edges and adding weight to the circumference until perfect balance is achieved.

Mount your rotor when balanced in a frame consisting of four 2 × 4's securely bolted at the corners with metal plates. Raise the structure at least 6 feet and preferably 10 feet above the ground, anchoring the rotor by guy wires and turnbuckles.

To transmit power to a pump the rotary motion of the wind machine must be converted to reciprocating motion. This can be

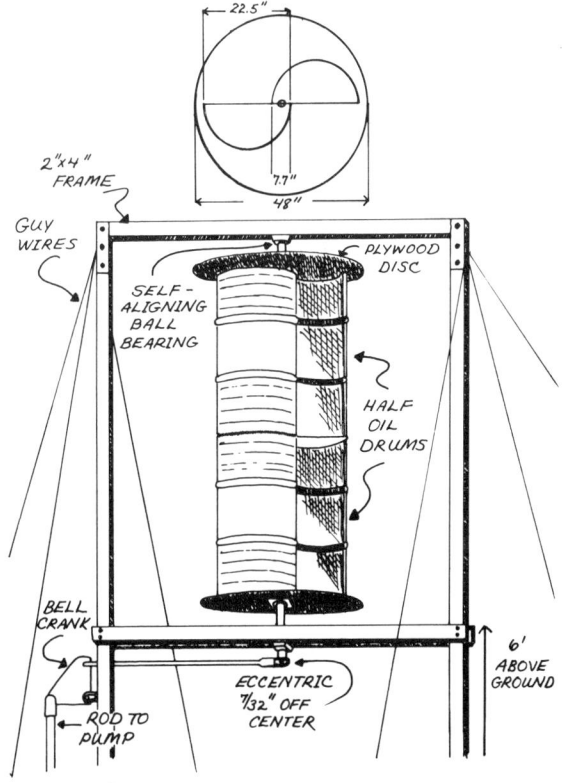

SAVONIUS ROTOR WIND MACHINE

FIGURE 37

done quite simply. Prepare an eccentric drive by fitting a steel cup to the lower end of the rotor shaft. Drill and tap a hole 7/32 inch off center and screw into it a bolt. This bolt must fit snugly into the big end of the connecting rod which can be made from the connecting rod of a motorcycle or small car engine. Cut the rod and weld on 6 inches of tubing of 1-inch inside diameter to receive a connecting rod made of wood 1 inch square. Attach a fork connection to the other end of the rod drilled to receive a pin ½ inch in diameter.

Connect the rod to the bell crank, which can be made from a piece of ¼-inch steel plate. Drill three holes in the plate after reinforcing the steel with welded bosses. If possible, line the holes with bronze bushings to increase the life of the pins and make arrangements for lubrication.

This wind machine can be attached to any piston or diaphragm pump, provided the pump stroke is not too long.

2. Propeller

The Savonius rotor, although easy to construct, is not the only way of converting air movement to power. The traditional windmill with its sails outspread gives more power than does the rotor. Anyone wishing to construct a wind machine of this type should bear in mind the following facts:

(1) Power output varies as the square of the propeller diameter. A propeller 6 feet across gives four times as much power as does one 3 feet across. But as size increases so do stresses, and the bigger the propeller, the stronger must be its construction. One of the biggest wind machines ever built (on Grandpa's Knob in Vermont) measured 175 feet. But despite the highly sophisticated design and great engineering know-how one of the blades of this monstrous propeller broke and the project was abandoned (see P. C. Putnam, *Power from the Wind*, Van Nostrand Co., New York, 1948).

(2) Number of blades and their design influence the speed at which the propeller turns. *Tip speed ratio* is the ratio of the speed of the blade tips to the speed of the wind. Propellers

designed to generate electricity usually have two or three blades and ratios between 5 and 10. Such blades do not start easily in a low wind. Propellers of the kind commonly used to pump water have as many as 20 blades and a tip speed ratio of 1 to 3. Such propellers have low starting torques and perform well in low wind velocities.

(3) The type of propeller best suited for pumping water (low tip speed ratio) is not the best for generating electricity.

For the pumping of water, wind power is excellent if a large reservoir is provided to hold the water pumped. As long as the reservoir is large enough there will be a supply of water during those periods when the wind does not blow. But for generating electricity wind power leaves much to be desired. First, it is intermittent. Second, it is not a very strong source of energy. "Like a good site for a large dam, a good site for a large block of wind power is a topographical rarity" (Putnam). The power which can be extracted from the wind is proportional to the *cube* of the wind velocity, which means that if you can get 1280 watts at a wind speed of 20 miles per hour you will only get 20 at a speed of 5 miles per hour. During calm spells you will not get any power at all.

It is out of the question to expect the wind to generate enough electricity to operate such energy-guzzling devices as electric space heaters, water heaters, clothes dryers, color televisions. No true Whole-earther would have such monstrosities on his place anyway. But even to power such legitimate devices as refrigerators and deep freezers will require a wind machine plus generator of substantial size. Nor will it be easy for the Whole-earther to build such a device himself. Used generators of the sort one might pick up cheaply are designed to run off diesel engines at speeds of from 1,500 to 3,600 rpm (revolutions per minute). A wind machine rarely runs faster than 300 rpm. A special type of generator is required for these low speeds.

At this point the true Whole-earther, unless he happens to be a mechanical genius, may be willing to let someone else build his wind machine. This is, of course, possible. If he wants a wind machine for pumping water he can get one from the Aeromotor Co. of Broken Arrow, Okla. 74012. If he needs a machine to generate modest amounts of electricity, enough to light his house for instance, he can get a Wincharger with a 200-watt generator from Dyna Technology, P.O. Box 3263, Sioux City, Iowa 51102 (around $400). For a more powerful machine (1,000 or 2,000 watts) he can apply to Pye Industries Sales Ltd., 2-22 Hargreaves St., Huntingdale, Victoria 3166, Australia; or Quirks, California Agent, P.O. Box A, Guerneville, Calif. 95446. An even more powerful machine (up to 5,000 watts) can be had from Elektro GmbH of Winterthur, Switzerland. This machine generates 110 volts direct current in conjunction with a bank of storage batteries. Needless to say, all wind machines generate direct current so that, if 120-volt alternating current appliances are to be used an inverter must be provided (which will steal 10 to 20% of the power).

TABLE VI
Wind machine output in watts

Wind Velocity, miles per hour	Diameter of propeller, feet							
	2	3	4	5	6	8	10	12
5	0.6	1	2	4	5	10	15	21
10	5	11	19	30	42	75	120	170
15	16	36	64	100	140	260	400	540
20	38	85	150	240	340	610	950	1,360
25	73	160	300	410	660	1,180	1,840	2,660

Purchase and installation of such relatively large wind machines should be preceded by a careful study of the frequency and strength of wind. Table VI gives the theoretical power available from high-performance wind machines in winds of different velocities.

Self-recording anemometers are available and careful studies should be made to ascertain whether a given site is really windy enough to warrant installation of a wind generator. All but the windiest regions are liable to periods of calm, therefore a careful Whole-earther will back up his wind generator with a second source of power (a gas engine generator, preferably powered by methane).

WATER POWER

Like wind power, water power is generated by the sun, which lifts water vapor from the ocean. The vapor is moved inland by the wind currents, condensed and precipitated in the form of rain or snow. By storing this water in a dam and leading it through a turbine to a lower level the water is forced to do work. The amount of work depends on the difference in height between the top of the water in the dam and the turbine, and on the amount of water flowing through the turbine.

Any Whole-earther on whose property there is a stream should consider the possibility of harnessing the stream flow. He can decide if this is worth doing by calculating the amount of electricity that his stream could generate if its energies were suitably directed. These calculations must be based on the stream flow or water level in a dam at the time of lowest water supply (late summer). If the water source dries up at this time a second source of power will have to be used to supplement the water.

To calculate the power a stream can generate one must know two things: the amount of water passing through the wheel in a minute, and the "head," or vertical distance, through which the water falls. We can express the first amount in cubic feet per minute (Q) and the second in feet (H). Then the power available at the site of the water wheel can be expressed in horsepower (HP) by using the following equation:

$$HP = \frac{62.4 \times Q \times H}{33,000}$$

There are several kinds of water wheels. The *impulse* wheel is activated by the pressure of a jet of water impinging against a set of specially shaped buckets on a rotor. The *overshot* wheel uses the weight of water in the buckets to carry the wheel around. The *undershot* wheel is activated by the force of the current beating against paddles projecting into the water. The *propeller reaction* wheel is moved by current flow acting on a propeller-shaped runner. Choice of wheel type will depend on the values of the H and Q that can be obtained. For an impulse wheel a high value of H is necessary, for an overshot wheel a high value of Q, for an undershot wheel a rapid stream flow.

The first step, therefore, is to decide where to locate a dam and where to place the power house. A hilly place with a rapidly flowing stream is ideal for the installation of an impulse wheel. The dam need not be large but must have sufficient capacity to supply the wheel with the flow of water required to maintain a charge in the batteries. To estimate the flow required, consult table VII (taken from "Your Own Water-Power Plant," reprinted from *Popular Science Monthly* in *The Mother Earth News,* no. 14, 1972).

TABLE VII
Relation of head and flow rate to power produced by an impulse-type water wheel

Head, feet	Flow, cubic feet per second	RPM	Horsepower
25	0.43	350	1.0
30	0.51	390	1.3
40	0.59	450	2.0
50	0.66	500	2.8
60	0.73	550	3.75

The impulse-type wheel can be located in a power house quite distant from the dam. Its water supply can be led to it by means of plastic hose, which must be wide enough to deliver the necessary flow (about 4 inches) in the impulse wheel described here. The dam, therefore, should be made as high up on the hillside as possible and the pipe should lead water into the wheel with as few bends as possible to reduce loss of power through friction.

The overshot wheel must be located near the dam. It is suitable for locations in which large values of H cannot be obtained. A 5-foot wheel requires a head of only 6 feet 3 inches so any location in which this much fall can be obtained will be suitable.

A dam is necessary only to provide a reserve of water for the water wheel. In a fast-flowing mountain stream the dam can be quite small. Indeed, because dam making is expensive in time, labor and materials, the dam should always be as small as possible. The generator, if it is of adequate horsepower, will rarely have to be run 24 hours a day so the dam can be designed to store water when the generator is not in use. All dams holding back large volumes of water represent a hazard and great care must be taken with their construction.

The cheapest dam to build in terms of materials is the earth dam. It is desirable when starting such a dam to construct a seal of planking, plywood or masonry extending into the sides and bottom of the dam and reaching above water level. This prevents water from seeping through the dam and eroding away the earth before it has time to pack solid. Earth for the dam can best be obtained by excavating the area behind it. In all dam building the problem has to be faced of what to do with the water while constructing the dam. The best solution is to divert the stream. In any case, dam building should always be done in late summer when stream flow is minimum.

The most important part of an earth dam is the spillway. If this is poorly constructed, too short or too narrow, flood waters at times of heavy rain will overfill it and erode the earth around it. The spillway therefore must be wide enough to take flood waters. It should be lined with concrete and the walls should be made of stone or concrete block and it should extend far enough from the dam to prevent erosion. If the dam does not have to be more than 3 feet high a framed dam can be constructed of heavy timber and sealed on the downstream side with fill. A spillway of planking can be installed on such a dam.

Plans for the construction of an impulse wheel (fig. 38) were given in *Popular Science Monthly* and reprinted in *The Mother Earth News*, no. 14, 1972. It is to this wheel that the figures given

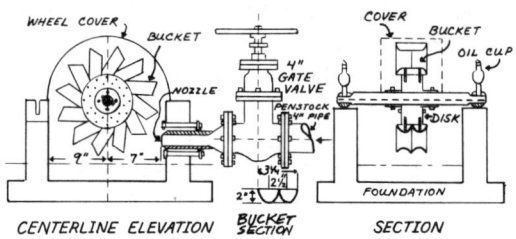

IMPULSE WATER WHEEL

FIGURE 38

in table VII refer. This form of water wheel is known as a Pelton wheel and has much in common with the Savonius wind rotor. The paired buckets should be mirror-smooth inside and must be easily removable for replacement. Above all, they must balance. Weld beads can be laid on the backs of any light buckets and ground smooth to minimize spray.

The shaft need not run in ball bearings, as the wheel rotates at a relatively low speed. If plain bearings are employed provision must be made for proper lubrication. The bearings are attached directly to the foundation, the function of which is to hold the wheel and nozzle in correct alignment. The outflow from the wheel must be conducted away from the power house in a concrete or wooden duct to prevent erosion.

In an overshot wheel such as the one shown in fig. 39 (also taken from the articles in *Popular Science Monthly*) the power developed is dependent on the size of the buckets and the diameter of the wheel. The wheel shown is 5 feet in diameter but its power can be raised from ½ to 1 horsepower by increasing the bucket width from 16 to 32 inches and the flume width from 13 to 29 inches. Before deciding on wheel size you will need to know the value of Q (flow rate of water). The value of H is fixed for this wheel at 6 feet 3 inches. To obtain 1 horsepower the value of Q would have to be about 100 cubic feet per minute.

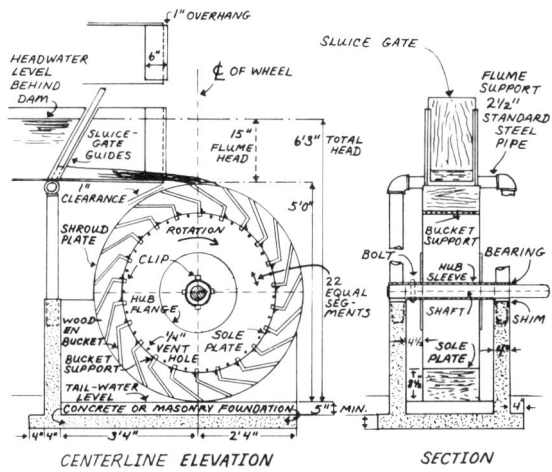

OVERSHOT WATER WHEEL

FIGURE 39

Water is led to the wheel in a wooden flume, which should be carefully bedded in concrete where it enters the dam. Rate of water flow is controlled by a sluice gate operated by a rack and pinion. The wheel itself can be constructed from ⅛-inch sheet steel. It consists of the following parts: (1) two shroud plates in the form of 5-foot discs; (2) a sole plate continuously welded to each shroud plate; (3) two ¼-inch steel hub flanges to which the shroud plates are continuously welded; (4) 22 buckets, which can be sheet steel or wood; (5) one shaft turned from 3-inch-diameter steel; (6) one hub sleeve of 3-inch steel pipe; (7) two bearings with renewable liners and oilers.

The wheel turns slowly (10 rpm) but is heavy and runs constantly, so good lubrication is essential. One who lacks welding equipment can build this wheel from marine plywood, protecting it with a layer of fiber glass and resin. The sole plate in this case should be made of 22 plywood segments attached to the shroud plate with metal clips and glued. The plywood plus fiber glass plus resin gives a very strong structure which is lighter than steel. An even lighter wheel can be made by sandwiching polyvinyl chloride foam between two layers of fiber glass mat (for details of techniques, write Boatex Fiberglass Co., 11 Tech Circle, Natick, Mass. 01760) using ¾-inch marine plywood for the hub flanges. Though no data is available on the behavior of these modern materials in water wheels their use in boats suggests that they could be adapted for this purpose.

After the wheel is in position, adjust the sluice so that the buckets turn one-fourth full, which will give the wheel a speed of 10 rpm. If the buckets run more than one-fourth full, efficiency will drop because centrifugal force will throw water from the buckets.

Those unwilling to build their own water wheels can purchase them ready-made from James Leffel and Co., Springfield, Ohio 45501. The company makes a series of hydroelectric power units in which a propeller-type turbine is connected via a governor to a generator which can be either alternating or direct current. This unit is supplied in sizes ranging from 0.5 to 10 kilowatts. Required heads range from 8 to 25 feet.

METHANE POWER

Of all power sources available to the rural commune methane power is probably the most reliable. Winds may fail, the sun may not shine and water power may not be available. But wherever there are animals, including human ones, there will be shit and where there is shit there can be methane.

Methane (CH_4) is formed from animal wastes by bacteria in two stages. First, the complex molecules are broken down to simple ones, sugars, alcohols and peptides. These accumulate and are acted on by a second group of bacteria to form either carbon dioxide or methane. Methane will be formed only if the fermenting takes place in the absence of air (anaerobically). The most suitable temperature for the reaction is 35° C (90–95° F) and the best pH is between 6.8 and 8.0.

Much research into methane generation has been carried out at the Gobar Gas Research Station at Ajitmal in northern India. India is in the situation now in which the fuel-guzzling, developed nations will find themselves in a few more decades. It has eliminated its forests and is reduced to using cow dung for fuel. Three-quarters of the country's annual billion tons of cow manure is burned for fuel, robbing the country's soil of badly needed nutrients. Obviously, it would be more hygienic and less wasteful to convert the cow dung to methane, use the methane for fuel and return the waste products of the fermentation to the land. To do this requires some know-how and this know-how has been accumulated largely by Dr. Ram Bux Singh, director of the Research Station. His books *Bio-Gas Plant* and *Some Experiments with Bio-Gas* are available from Gobar Gas Research Station, Ajitmal Etawah (U.P.), India. Much of this information has been made available in *The Mother Earth News,* nos. 3, 12, 18.

It is possible to brew methane in any airtight container, provided one arranges to remove the gas. That ingenious inventor Harold Bate uses nothing more elaborate than an old domestic water heater which he fills with the manure extract and seals (see *The Mother Earth News,* no. 10, 1971). Bate recommends a half-and-half mixture of chicken and pig shit mixed with straw (3:1). He finds it advisable to stack the mixture, douse it with water and let it ferment about a week before filling the digester. The manure can be stacked around the digester to provide heat. The methane reaction goes best at 85–90° F.

A sealed digester of this type will generate gas under pressure (fig. 40). This gas should be removed continuously through a narrow tube fitted with a flame arrester in the form of a piece of wire mesh. Always remember that a mixture of methane and air is explosive so the first gas brewed in the generator is apt to be dangerous. It should be allowed to escape. Once the methane

272 ENERGY

has displaced the air the gas can be safely stored. A sealed digester, like a domestic pressure cooker, should have attached to it a pressure gauge. The Harold Bate digester has a safety valve set to blow at 60 pounds per square inch.

Sealed containers such as Harold Bate uses must be charged and emptied at intervals. His will hold 300 pounds of manure and will generate 1,500 cubic feet of methane. A much larger digester can be designed capable of digesting 100 pounds of manure

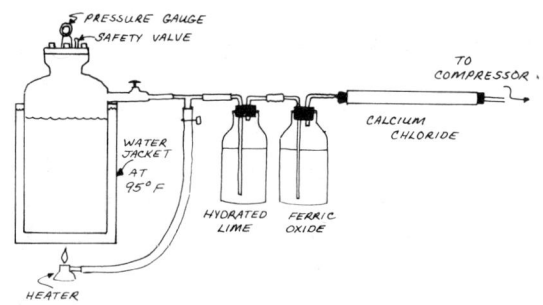

CLOSED SYSTEM GENERATOR

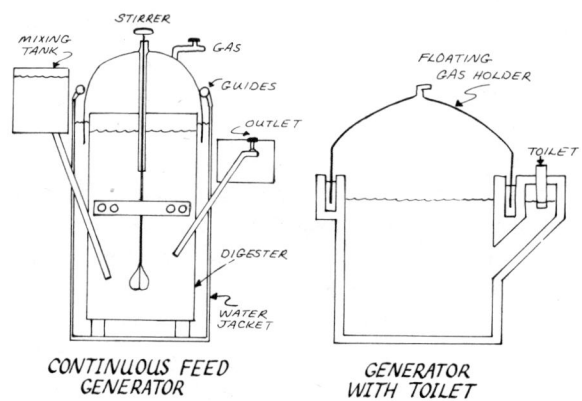

CONTINUOUS FEED GENERATOR   GENERATOR WITH TOILET

METHANE GENERATORS

FIGURE 40

every 24 hours (*The Mother Earth News*, no. 12, 1971). The construction of this rather large unit will be justified only if the Whole-earther planning to use it can keep it busy. It devours the fecal products of five cows per day, which means that, to satisfy its demands, one must either keep one's cows indoors or chase after them and gather the shit as they deposit it.

The combined digester and gas collector need not be as large as the one shown. It can be scaled down to suit the raw materials supply. Several designs are possible, including one which provides the Whole-earther with an opportunity to add his own feces to those of his animals (fig. 40). The digestion of the waste material will be more complete if some device is used to stir the mixture at intervals. Provided it is supported in a water bath, the digester does not have to be rigid; indeed a very satisfactory one can be made of large-diameter plastic tubing along which the waste can be moved by periodic agitation.

The heating quality of the gas developed in the digester will be much improved if the carbon dioxide is removed. This can be done by bubbling the gas through a slurry of hydrated lime. The lime traps the carbon dioxide and is converted from calcium hydrate to calcium carbonate. Fresh calcium hydrate must be added from time to time as the slurry becomes converted. To determine whether it is working efficiently, prepare a slurry from fresh hydrated lime, allow it to stand and remove the clear supernatant (lime water). If the gas from your digester produces a milky precipitate on being bubbled through the lime water it still contains carbon dioxide, which means that it is either passing through the slurry too quickly or that the slurry is spent. Spent slurry should be used on the garden. It is a source of calcium.

Another gas that may contaminate your methane is hydrogen sulfide (H_2S), which imparts an odor of rotten eggs to the gas. It must be eliminated if you intend to use methane to run an internal combustion engine. Hydrogen sulfide can easily be removed by passing the gas through a slurry of ferric oxide or through a tube of moist iron filings.

If you wish to use the methane solely for heating, the pressure generated by the weight of the gas holder should not be more than 10 to 20 grams per square centimeter. It is easy to calculate the pressure if you know the weight of the floating gas holder. For example, a gas holder weighing 300 kilograms having a cross-sectional area of 2.54 square meters would exert a pressure of $300/2.54 = 118$ kg/m^2 $= 11.8$ g/cm^2 $= 0.17$ pound per square inch. This is about right. A heavy concrete gas holder would have to be counterweighted to reduce pressure. A light gas holder of, say, plastic would have to have weight added. *Always keep a positive pressure on the gas.* Negative pressure can result in air being sucked into the system, which can lead to an explosion.

If you wish to make your methane portable you can use a compressor. Harold Bate compresses the gas to 1,100 pounds per square inch. The gas at this pressure must be contained in a solid pressed steel tank equipped with an excess pressure relief valve. To transfer the gas to the engine of a car several special items are required whether you use compressed methane or the liquid propane (LP) now being widely used in place of gasoline. Propane (C_3H_8) is a more efficient fuel than methane because it yields more heat per molecule. It is, however, derived from the same source as gasoline and, for this reason, apt to become scarce as the world's oil reserves become depleted.

To run a car or truck on methane or liquid propane five items are necessary (see "Convert Your Car to Propane" by Jerry Friedburg, *The Mother Earth News,* no. 15, May 1972):

(1) Fuel tank. A regulation motor vehicle LP tank will hold 14 gallons, has a gauge and necessary safety features. (2) Converter. This device uses engine heat to convert liquid propane to a gas. It supplies just as much of the fuel as the power plant demands. (3) Spud-in jet. This jet introduces the fuel into the carburetor. (4) Hoses. Hose A takes the liquid propane or methane to the converter and has to withstand pressure. Hose B carries the fuel from the converter to the carburetor. (5) Fittings and hardware. A foot or so of heater hose and two Y fittings are needed to conduct heat from a water-cooled engine to the converter. In an air-cooled engine (as in a Volkswagen) this is accomplished with

copper tubing of ⅜-inch outside diameter. A fitting is necessary to adapt hose A to the converter's LP input. Another fitting is needed to connect the converter's vapor output to standard fuel line of 5/16-inch outside diameter.

All this equipment except the tank is available from Jerry Friedburg, Arrakis Volkswagen, Box 531, Point Arena, Calif. 95468. The 1972 price was $65 for a Volkswagen, $60 for other engines under 150 horsepower and $70 for engines over 150 horsepower.

V CRAFTS

One of the essential characteristics of the true Whole-earther is a desire to escape from the plastic age and return to the days when household objects were produced by craftsmen with love and care and a long tradition behind them. Such craftsmen are vanishing. If you doubt it, read *Traditional Country Craftsmen* (by J. Geraint Jenkins, Routledge and Kegan Paul, London, 1958). Again and again the phrase recurs, "this craft has almost completely disappeared from areas of rural England in which it existed for centuries."

The same thing has happened in America. Only in more or less isolated areas in the Appalachians are the old crafts practiced. Thanks to the splendid efforts of Eliot Wigginton and his students much of this know-how has been preserved in *The Foxfire Book* (Doubleday, New York, 1972).

The true Whole-earther can master these crafts and practice them not so much to earn a living as to enjoy a form of creative activity that can pleasantly occupy winter days when nothing else can be done.

BASKET MAKING

This craft is one of the very few that has not been taken over by the machine. There are so many kinds of baskets and they come in so many different shapes and sizes that the machines that are rendering the craftsman's skill obsolete cannot compete in this area.

1. Split oak baskets

Lengths of split oak are used to make a variety of sturdy baskets. Oak splits are probably the most readily available basket making material and the craft is not very difficult (see *Foxfire Book,* pp. 115–127). The splits are made from a white oak sapling 4 to 6 inches in diameter. The trunk must be straight, free from limbs, knots and imperfections, and should be 6 to 9 feet long. The American basket makers gather the timber in summer or fall and try to make the splits on the day they cut it. The British trug maker uses white pollard willow for boards and ash or chestnut for the frame.

Making oak splits calls for experience. Starting at the small end of the trunk a froe is driven in with a mallet. Wooden wedges are used to carry the split down the trunk and the two halves split into quarters. The bark is shaved off and each quarter is split into eighths with a sharp steel wedge. Beginning at the butt end of one of the eighths it is split in half parallel to the grain of the wood. The split can be started with a pocket knife and continued with the hands, but the perfect tool is the British spar maker's billhook. The split tends to "run out" and only by experience can one learn to avoid this.

The splits can be used for bottoming chairs or making baskets. Baskets of splits can be of two types: trugs or hampers. A trug is based on two hoops shaped from split oak or ash and secured with nails (fig. 41). The splits are woven around the hoops, beginning where they intersect. Extra ribs are added as required. Ten ribs are needed half an inch wide, pointed at both ends. The splits are woven in and out of the ribs. In a Sussex trug the cleft

278 CRAFTS

willow boards are nailed to the frame after being bent in a special setting brake.

Another basket made from oak splits is the hamper. A hamper does not need a frame. It is woven entirely from splits. The first ribs are tacked lightly to a board to hold them in place. The bottom of the hamper is woven, then the splits are moistened and carefully bent upward. Then pliable splits are woven in and out of the ribs to make the sides. The splits must be kept close together and tight. At the top the ribs are cut off even and secured

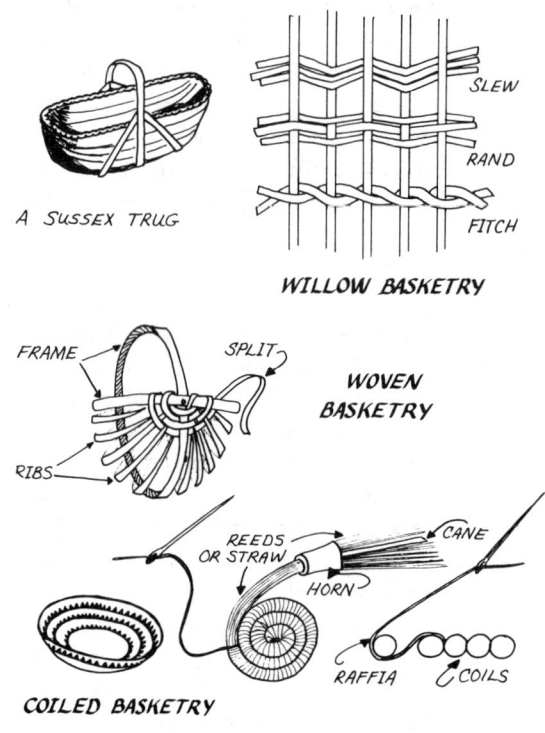

BASKET MAKING

FIGURE 41

with a split on either side. These splits can be stapled to the ribs or held in place with a cord.

2. Osier baskets

Osier basket making is one of the oldest crafts and takes the craftsman clear back to the Neolithic. Any rural commune having a patch of fertile but swampy ground or a year-round creek can engage in this craft using its own home-grown willows. In England willows for basket making are grown mostly in a part of Somerset known as Sedgemoor. The soil is prepared by deep plowing or by digging to a depth of two *spits* (spade thrusts). The willow sets are purchased in the autumn, kept in damp grass until late winter. Sets are cut 12 to 15 inches long and pushed into the soil until about 7 inches of set appears above the soil. Spacing is 15 inches between sets, 22 inches between rows. The most popular varieties are Black Mawl or Champion Rod.

Once planted, an osier bed may be harvested for as long as 80 years. It is therefore worthwhile to prepare it carefully. The crop is harvested by cutting the rods as near the stool or stump as possible. They are cut with a quick upward movement of a sharp billhook and the rods are tied in bundles. After 3 years the bed will bear a full crop and will generally yield 2–3 tons of withy per acre.

Willow rods are generally treated before being used for basket making. *Brown willow* is prepared by steaming the rods in a steaming chest, then stacking them in the open to dry for several weeks. It is used for making wickerwork garden furniture and items such as watercress baskets, which are exposed to damp. *Buff willow* is made by boiling willow wands vigorously for 5 hours and leaving them in the water for 24 hours or longer. The tannin in the bark stains the rods a golden brown. After removal from the boiler the willow rods are stacked in heaps, kept moist and later stripped of the adhering bark. *White willow* can be prepared only from rods harvested in autumn or winter and placed in special pits with a few inches of running water. In spring, when the sap rises in the rods, the bark is stripped by passing each rod through a forked instrument called a brake.

The basket maker prepares for the day's work by putting willow reeds in water to soak overnight. With these moistened rods in easy reach he seats himself cross-legged at a lap board and begins to weave. For a rectangular basket he fixes a number of rods in a small wooden clamp and weaves them together with smaller rods until he has a firm base. For a circular basket he places a group of rods (generally four) across another similar group of rods and binds them together. He continues to weave by opening out the ribs until they radiate from the center like the spokes of a wheel. More stakes are added as required. When the base is complete the edge is reinforced with a stouter weave and extra rods are inserted to form the side. A willow hoop passed round the upright stakes keeps them in place as the work proceeds. The long ends of the stakes are woven along the rim to form a border.

Willow basket making requires both strength and dexterity. Even after soaking in water the material is hard to bend and weave. The basket maker requires a few simple tools. His *lap board* can be a sheet of plywood supported at a convenient height by short legs. His *bodkin* can be of iron or bone. It is used to make openings in the weave for the insertion of rods. His *beating iron* is a piece of iron 9×3 inches used for beating the weave into place. The *picking knife* is short and broad-bladed, used for trimming the finished basket. The *commander,* an iron rod ringed at the tip, is used for straightening the stakes and beating the weave. A tank for soaking the willow and a heavy weight to hold the basket in place during weaving are also necessary.

A skilled basket maker uses several kinds of weave. *Randing* involves working a single rod in front of and behind the stakes. *Slewing* is like randing but two or more rods are worked at a time. *Fitching* involves working the rods alternately over and under each other. *Waling* involves working three or more rods alternately one by one in front of two, three or more stakes and behind one. This weave may be used to edge the bottom of the basket (fig. 41).

3. Coiled baskets

The coiled basket is made by using a bundle of fairly stiff material as a core and binding it in the form of a spiral with a second, more flexible material. The core material can be split willow, yucca, cane or straw. A decorative and strong coiled basket can be made from cane and raffia. Cane is made from the pith of the tropical rattan palm and is very flexible and easy to work. The cane provides the spiral coil and the raffia, threaded through a needle, is used to sew the spiral together. A figure-eight stitch is used (fig. 41).

To make a large coiled basket it is necessary to thicken the coils. This can be done with straw. Cane can be used as the core of the coil and lengths of straw or reed placed around it to form a coil as much as an inch thick. The thicker the coil, the more rapidly the basket will be completed. A useful way of holding the coil together is to lead it through a piece of cow's horn cut to hold a bundle of the right diameter.

The coil method of basket making was used in England, Scotland and Scandinavia to form objects of various shapes ranging from cradles to beehives (skeps). Straw of winter wheat that had not been threshed, bent or bruised was used for this purpose and the binding material consisted of long shoots of bramble cut during the winter, split in four sections and smoothed with a knife. The tools consisted simply of a sharp pocket knife, a section of cow horn to hold the bundle of straw and an awl made from a piece of bone from a horse's hind leg. The section of cow horn was ¾ to 1 inch in diameter. A small cylinder of leather could be substituted for the cow horn.

POTTERY

Along with spinning and handweaving, pottery is one of the more popular crafts among Whole-earthers eager to establish contacts with basic materials. Clay is part of the fabric of human civilization. He who shapes vessels in clay joins a long line of craftsmen whose art goes back to the Neolithic. Whether one works clay by the slab method, builds it in coils, pours it in molds or spins it on a wheel the operation is profoundly satisfying.

Obviously the potter's first task is to obtain clay. He can do this very easily by going to the nearest hobby shop and buying it. Or he can go and dig his own clay, in which case he may end up with a material that is either impossible to work, cracks on drying or is impossible to fire. Any Whole-earther wishing to use a local clay would do well to try various additives with a view to improving its behavior. The clay should be removed from a level about 2 or 3 feet below the surface to avoid contamination with top soil. Allow the sample to dry, break it into small pieces, sprinkle with water. When it becomes soft enough to work, form it into a block $5 \times 1 \times \frac{1}{2}$ inches. Allow it to dry. It should show no sign of cracking. It will, of course, shrink in size but if the 5 inches has been reduced to less than $4\frac{1}{2}$ inches the shrinkage may be too great. The condition can probably be remedied by the addition of sand, flint or grog. Addition of fire clay may also be helpful.

If, instead of using local clay, the Whole-earther wishes to purchase this material he is apt to be confused by the many varieties available. He can avoid this confusion if he simply remembers three categories: *Earthenware* is the kind of clay from which flower pots are made. It is coarse, has a low resistance to chipping, is suitable for the formation of rather thick-walled products and can be fired at a rather low temperature. *Stoneware* has a finer structure than earthenware and fires to a tan or gray color rather than red. It is suitable for kitchen and baking ware. *Porcelain,* which is the aristocrat of clays, has an extremely fine grain, fires white at a high temperature and can be used for the creation of delicate, thin-walled products.

Before clay can be used it must be wedged. This process involves a wedging table onto which a piece of clay can be thrown. Each time after throwing, the clay lump is cut in half with a piece of piano wire stretched across the table. The two halves of the lump are rejoined and thrown down again. This process gradually forces air pockets out of the clay until the lump, on cutting, is seen to be completely free of air. Once a lump has been wedged it must be wrapped in cloth or plastic and stored in a closed container until used.

Preparation of slip involves adding enough water to clay to make a suspension having the consistency of thick cream. Slip must be free of air bubbles and lumps. It should be strained through a nylon stocking and allowed to stand in a jug or other container until air bubbles rise to the top. To avoid introducing bubbles slip must be stirred slowly.

Clay that is too wet to work can be easily dried by the use of plaster bats. Such bats can be made from potter's plaster. For a 10 × 10-inch bat, add 5½ pounds plaster to 2 quarts of water and pour into a cardboard box. Circular bats can be made in pie plates but the walls must be lined with a layer of clay to facilitate removal. Pie-sized bats should be at least an inch thick and several should be prepared and allowed to dry.

Potters use various terms to describe the condition of their clay. *Slip* or *engobe* is a liquid suspension of clay that may be thick or thin and is used for decoration, for casting in molds or for sticking clay together. *Slurry* is a very thick slip. *Plastic* is clay that is wet enough to be modeled, and the starting material for most work with clay. *Leather-hard* clay is partly dried plastic. It can be cut or carved or stuck together with slip, will bend but not easily break. *Greenware* is clay that has been air-dried, is brittle and breaks easily. Greenware can always be converted back into workable clay by addition of water but once the greenware has been fired it changes into a different substance. This is called *biscuit* or *bisque* and results from exposing greenware to temperatures of 1,000° C or over. Biscuit ware can be coated with a glaze and fired again at the temperature needed to melt the glaze. This is the final stage of the potter's art and the product is known as *glost*.

1. Working clay

There are several ways of working clay that do not require a potter's wheel.

Slabs are made by rolling out a piece of clay on a sheet of canvas or burlap. The rolling can be done with a rolling pin. To

284 CRAFTS

control the thickness of the sheet two strips of wood are placed one on either side of the piece of clay. To form a vessel the four sides and bottom can be cut out, using a paper pattern, and the edges of the pieces stuck together with slip (fig. 42). Once the clay is leather-hard the piece can be smoothed and shaped without danger of its disintegrating.

Shallow dishes and similar pieces can be shaped in a plaster mold simply by lifting the slab on its canvas and letting it rest against the mold held almost vertically. Lower the mold gradually and peel off the supporting cloth. Carefully lift the edges of the

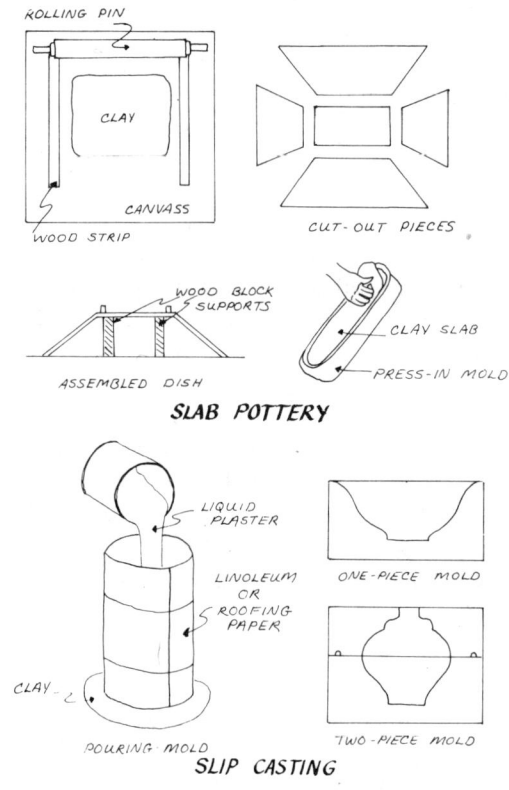

FIGURE 42

clay until the clay slab subsides into the mold without any air bubbles. Use the bowl of a wet spoon as a smoother to make sure all the air has been driven out. Trim off waste clay at the edge. Slabs can also be shaped on convex molds.

Coils are made by rolling clay on the bench with both hands. The bench and hands should be moist and the clay rolled quickly and uniformly. A number of such rolls of equal thickness must be prepared. To make a coil pot, first roll out a slab for the bottom. Mark a circle on the slab and cut away the excess clay. On the circular base, build up the sides of the pot with coils, cutting the roll to the right length, welding it to the base. Place each additional coil on top of the one below, welding ends and joints on both inside and outside. Allow coils to stiffen slightly before adding new ones, otherwise the piece is apt to sag.

For making coil pots of circular cross section it is handy to have a bench whirler or banding wheel, which is simply a turntable that enables work to be rotated. The pot should not be built directly on the wheel but on a plaster bat held in place on the banding wheel by lumps of clay. To ensure symmetry in your pot, prepare a template by cutting the outline of one side of the vessel in a piece of cardboard and placing it beside the coiled shape as coils are added.

If you wish to produce several pieces of the same design this can be done by means of a plaster mold. For open convex shapes like dishes or bowls a simple one-piece mold can be constructed (fig. 42). For more complex designs a two-piece or even a three-piece mold must be made. Lubricate the object to be cast with soap so that the plaster will not stick to it. Place the object on a piece of rolled-out clay, build a form around it with a strip of linoleum or heavy roofing paper. Prepare plaster by adding dry plaster to a volume of water sufficient to fill the mold. Enough plaster should be added without shaking or stirring to make a cone 2 inches above the water. Carefully stir it with the hand and then pour it into the mold, avoiding air bubbles. After the plaster has set, remove the object that was used as the model and let the plaster dry for a week.

Place the dry mold on a level surface and fill with slip, adding more slip as the level sinks. The dry plaster will absorb water from the slip, leaving a layer of thickened clay adhering to the mold. When this layer has attained a thickness of about ¼ inch, pour off the liquid slip and invert the mold. Let it drain for a half hour over a pan, right the mold and clean the rim. The casting will dry and harden, pulling away from the mold as it does so. When leather-hard it can be removed and allowed to dry completely.

Exactly the same method is used for two- or three-piece molds, slip being poured in and out through a hole in the top.

The above methods, especially that using the bench whirler for building pots by the coil method, will enable the aspiring potter to get the feel of the material. He will then be in a better position to embark on the most difficult aspect of the potter's art, namely throwing pots on a wheel. This involves making or buying a wheel, which can either be operated by the foot (kick wheel) or by an electric motor (fig. 43). The momentum of a kick wheel can be increased by using an old car tire filled with cement below the plywood disc. On the top of the wheel removable plaster bats can be used.

No amount of description will enable a would-be potter to master the art of throwing pots on the wheel. It is one of those skills that can be perfected only by practice. The following procedure is involved:

(1) Place a plaster bat on the wheel head and sponge it with water. (2) Wedge a ball of clay of suitable size, start the wheel revolving with the right side moving away from you. (3) Throw the clay ball on the wheel as near center as possible, wet hands and gradually force the clay ball to the center. (4) Bring the clay up in the form of a cone, then press it down again. Repeat the process until all irregularities in the clay disappear. This is known as *mastering* the clay. (5) Form the clay into a low cylinder and gradually bring up the walls of the clay between thumb and forefinger. Lubricate with water only when necessary.

Once the potter has mastered the art of forming a cylinder he can learn to modify this basic shape in any way he wishes by forcing the walls outward or inward. Once the piece has been shaped, lift off the whole bat with the pot attached and set it aside to dry. If the walls are too thick or too rough the piece can be turned or shaved at the leather-hard stage.

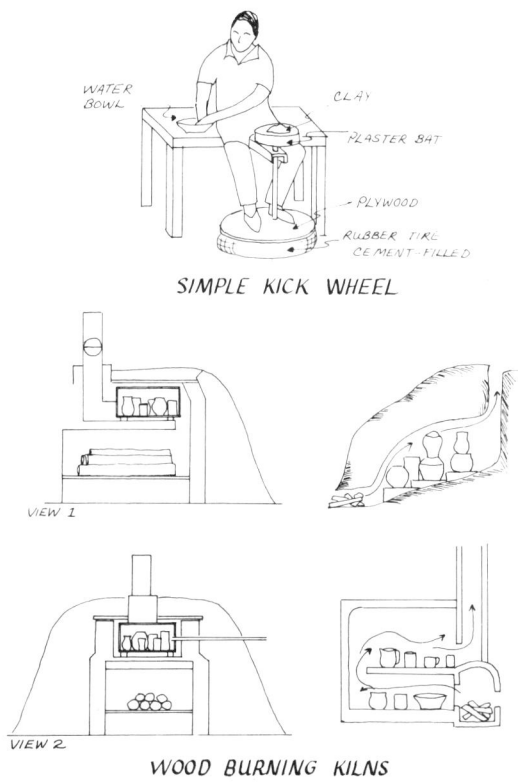

SIMPLE KICK WHEEL

WOOD BURNING KILNS

FIGURE 43

2. Decorating

Pottery can be decorated with slip or it can be glazed. Slip is applied to leather-hard clay with either a brush or a syringe. Banding or striping can be done on a bench whirler. *Sgraffito* involves covering the soft clay surface with a coat of slip of contrasting color, allowing the piece to become partly dry and cutting a design through the slip to the clay body beneath.

Glazing involves covering the body of the clay with a thin coat of glass. This makes the clay non-porous, improves its appearance and durability. A variety of glazes can be purchased ready-made. Colors are contributed by various metallic oxides, mainly cobalt (blue), copper (green), manganese (brown), chromium (green, brown or pink), antimony (yellow), uranium (red), iron (yellow). A clear lead glaze, using white lead, is easy to use but lead should be avoided for vessels that are likely to be used for food. Glaze is applied to biscuit ware, being either sprayed on, poured on or applied with a brush.

3. Firing

A pottery with no kiln is only half a pottery. Kilns can be built very simply and do not have to be electric. "When outdoor space is available," writes John W. Dougherty (*Pottery Made Easy*, Bruce Publishing Co., New York, 1939), "great experience and fun can be had building a kiln which uses wood or coal as fuel." Such a kiln can be built following the plans shown in fig. 43. If fire brick is available it should be used for the interior walls. The joints should be thin and made of three parts fire clay, one part sand and just enough water to make it spreadable. Common brick or rocks and ordinary mortar can be used for the exterior and a layer of earth over the kiln will greatly reduce heat loss.

Packing greenware in a kiln for biscuit firing involves the use of shelves and supports which will enable the maximum number of pieces to be fired at once. Greenware is very fragile and must be packed with care. The pieces, however, can be allowed to touch each other. For glost firing, pieces must be at least ¼ inch away from each other and must be supported on stilts or placed on shelves previously washed with flint wash.

The temperature of the kiln must be increased slowly. For biscuit firing a maximum of 950–1,000° C in 8 to 10 hours is fast enough. Standard cones which bend when the correct temperature has been reached should be placed in the kiln in such a position as to be visible through the spy hole. Three of these cones are used, the first bending below the required temperature, the second at the required temperature, the third above it. Biscuit is glazed at between 1,050° C (earthenware) and 1,250° C (stoneware). After the required temperature has been reached the kiln must be allowed to cool slowly. If the kiln is opened too soon, cracking will result.

BEADWORK

The art of working with beads is very old. In ancient Egypt it attained a high level of development. Coral, turquoise, glass and other materials were used and many of these decorative pieces have been recovered undamaged from Egyptian tombs. The American Indians used beads for money. Their "wampum" was prepared from clam shells, abalone shells or turquoise, each bead being carefully drilled by hand. Glass beads were introduced by the whites and quickly became incorporated into the fabric of Indian art. They were used to decorate buckskin skirts, belts, carrying bags or woven into headbands or collars.

Beads can be woven on a loom, sewn onto canvas or leather or "diagonally woven" without a loom. Loom weaving is probably the simplest method. It is possible to buy small looms specially designed for beadwork but they are generally narrow and suitable only for belts. The worker wishing to make larger objects can prepare his own loom. This may consist of nothing more than a pair of supports, a roller at one end and a nail at the other. Small notches ⅛ inch apart along the supports will hold the warp threads in position. For these threads dental floss is suitable or one can use linen thread, which should be waxed by pulling it through a lump of beeswax. The warp threads should be measured and be a few inches more than the length chosen for the work. If the work is to be 60 inches long the warp threads should be about 68 inches. There must be one more warp thread than the number of beads in each row so that there will be a thread on each side of each bead. The warp threads are tied together at one end and slipped over the nail. They are passed through the notches on the supports and attached to suitably spaced nails on the roller.

The warp threads can be tightened with the roller and the work started at the end farthest from the roller. A #11 needle threaded with #90 linen thread is passed through the warp threads from left to right and one end tied to the warp thread on the left, the loom being oriented so that the roller faces away from the operator. The needle is then passed through the beads from right to left in the correct order as determined by the pattern. The size of beads used will be determined by the degree of detail required, the smaller the bead the more intricate the pattern. If the warp threads are spaced ⅛ inch apart the best beads to use are size 4–0.

The warp threads must be opened to form a shed and the string of beads introduced into the shed as in ordinary weaving. Beads must be arranged in such a way that each has a warp thread on either side. The shed is then closed and the opposite shed opened. The beads are once more threaded, this time from left to right and again passed into the shed. When more weft thread is required it must be attached by means of a proper weaver's knot carefully adjusted to come at the edge of the work.

Though woven work of this kind will stand on its own it will be more durable and show to better advantage if backed by linen or soft leather.

Diagonal weaving does not require a loom, but is tedious to do because each bead must be added separately. The texture of the work is like brickwork, with the beads alternating, one up, one down. To start the work a bead is threaded on white sewing silk with a #12 needle. The silk is tied to the bead and other beads are added, the number being determined by the width of the work. If the work is four beads wide, thread the four beads, then pass the needle through bead 5 and line it up above bead 4 (fig. 44). Next pass the thread through bead 3, then through

BEADWORK 291

bead 6, then through bead 1. Turn the work over and pass the thread through bead 7, which will then line up with bead 1. The resultant work will be built up in staggered rows as opposed to the straight rows obtained when beads are woven on a loom. This staggering of the rows will, of course, affect the design. Though exceedingly tedious when small beads are used the diagonal method is quite satisfactory with large beads and can also be adapted for use with beads of many different sizes. It was popular among the ancient Egyptians and some rather attractive samples of this work have been recovered from Egyptian tombs

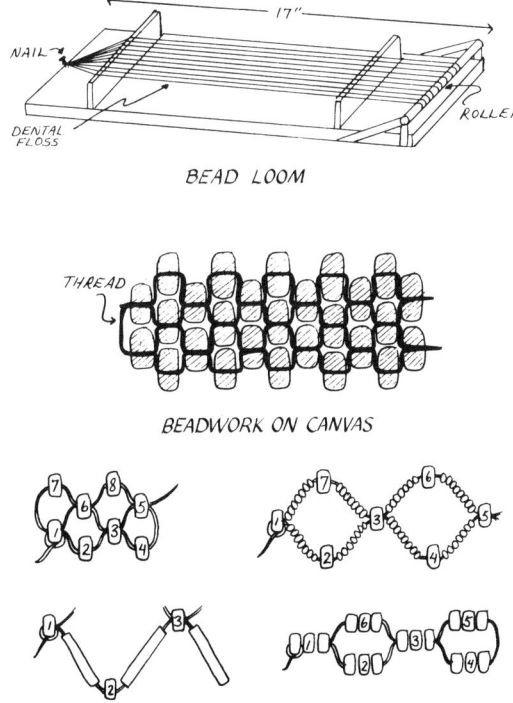

BEAD LOOM

BEADWORK ON CANVAS

DIAGONAL BEADWORK

FIGURE 44

(see M. White, *How to Do Beadwork*, Dover Publications Inc., New York, 1972).

The technique of diagonal weaving can be used to create various kinds of open beadwork on which the craftsman can exercise his/her ingenuity. All that one has to do is to interpose strung chains between the link beads. A simple example is shown in fig. 44, in which large link beads are used alternating with smaller chain beads. Cylindrical beads can also be used by this method. Further variants can be introduced by threading two or three beads at a time.

In addition to these techniques beads may be worked on a base of canvas or buckskin. This work is not difficult but it is slow as every bead must be sewn in place separately. Size 5–0 beads are suitable, and letter A sewing silk with a #12 needle should be used. The silk should be well waxed with beeswax. The needle is run for a short distance through three or four meshes on the canvas on the wrong side. A back stitch is made and the needle brought through to the right side. A bead is threaded and the needle passed through the horizontal meshes of the next square. This makes a diagonal stitch which holds the bead securely in place. All the beads must start the same way so the work must be turned upside down when the next row is sewn.

WATERCRAFT

Any Whole-earther who has access to rivers, lakes or the ocean may wish to attempt to build a boat. To do this, using only the materials readily available in the immediate neighborhood may tax his ingenuity. It is, however, quite possible, without using products of contemporary technology (plywood, fiber glass, resin, etc.), to build a good boat very cheaply if one is willing to follow certain traditional designs.

Native watercraft can be divided into six categories: (1) Rafts, much popularized by the Kon Tiki Expedition, were widely used by the natives on the South America seacoast. (2) Reed boats have been used since prehistoric times on the Nile, also on Lake

Titicaca and along the coast of Peru. (3) Kayaks are used by the Eskimos of North America and Greenland. (4) Open skin boats, of which the British *coracle,* the Irish *curragh* and the Eskimo *umiak* are examples, are still in use. (5) Bark canoes were used by the North American Indians. (6) Dugout canoes are widely used in Polynesia and elsewhere.

Making these craft demands varying degrees of skill. Anyone having access to a few boards can put together a raft. But to build a really good kayak or birch bark canoe or even a reasonably seaworthy reed boat is a fine challenge to the skill of the craftsman.

1. Coracles

The coracle is one of the most ancient and most simple of these native craft. The frame of a coracle consists of nothing more than a loosely woven willow basket over which some waterproof cover (hides in the old days, unbleached calico more recently) is stretched.

The frame of a coracle is made of willow rods cut from suitable willow trees in the autumn or winter. The rods are carefully split with a billhook. One coracle requires 10 longitudinal laths 7 feet 6 inches long and nine cross laths 6 feet 6 inches long. The shaved laths, smoothed with a drawknife, are immersed in a tub of boiling water until they are pliable. The longer laths are spaced 5 inches apart on a flat surface and lightly tacked to a board. The nine shorter laths are interwoven at right angles. Heavy weights are placed at the intersections of the laths and the sides are bent upward. Thin hazel saplings are woven to make the upper rim of the coracle, forming the gunwale. The seat, a piece of deal planking, is inserted in the gunwale (fig. 45).

The skin of the coracle, in the form of 5 yards of unbleached calico, is firmly strained over the frame and tacked in place. A series of coats of boiled linseed oil and pitch are painted on the fabric until it is completely waterproof.

No one would claim that the coracle is either safe, speedy or convenient. Being shaped like a shopping basket it tends to rotate if paddled by any but an expert and the high level of the seat makes it prone to capsize. It is a river craft, for quiet waters only, and was used to net salmon, the net being suspended between two coracles. "So efficient were the coracle fishermen in catching salmon that the restocking of rivers became a very serious problem and, throughout Britain, fishery boards drew up a great deal of legislation banning the use of the coracle as a fishing vessel" (J. G. Jenkins, *The Traditional Country Craftsmen*, p. 97).

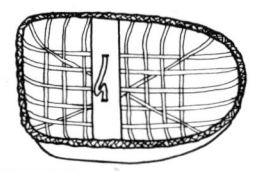

FRAME OF CORACLE

NUNIVAK ISLAND KAYAK LENGTH 15' BEAM 30" DEPTH 15 3/4"

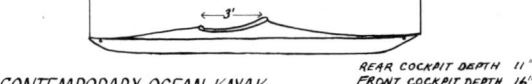

CONTEMPORARY OCEAN KAYAK REAR COCKPIT DEPTH 11" FRONT COCKPIT DEPTH 16"

"EL CABALLITO" REED BOAT OF THE PERUVIAN FISHERMEN

WATERCRAFT

FIGURE 45

2. Kayaks

The defects of the coracle can be remedied if the frame is modified to produce a vessel having the shape of a kayak. Any Whole-earther who needs a cheap but light vessel for fishing can build such a boat and learn a lot in the process. Its design will depend on the conditions under which it is to be used. The genuine Eskimo kayak is designed to fit the user so that it becomes almost an extension of his body. The Nunivak Island kayak is just 15 feet long, 30 inches in the beam and 15¾ inches at its maximum depth. This is one of the safest of the Eskimo kayaks, several of which are illustrated in *The Bark Canoes and Skin Boats of North America* (by E. T. Adney and H. I. Chapelle, Smithsonian Institution, Washington, D.C., 1964). Compared with the Greenland kayak the Nunivak Island version is rather dumpy and awkward but the beautiful lines of the Greenland kayaks result from their very narrow beam (as little as 19¼ inches in some cases). This makes them very liable to capsize, which is no disadvantage if one is expert with the Eskimo roll but may mean death by drowning if one is not.

Being myself an experienced kayaker I can say that a beam of 36 inches is very comforting in rough water, that a length of 15 feet is fine for flotation but that 12 feet is adequate if one wants a light boat. A depth of 11 inches is sufficient at the seat, and of 16 inches at the point of the cockpit. If the cockpit is extended forward to a length of about 36 inches an opening of sufficient size is obtained to enable the kayaker to extricate his fishing pole, tackle box, lunch, etc., without being a contortionist. Such a relatively large opening will prevent him from sealing himself into the kayak Eskimo-style, but a removable spray cover will make the boat fairly waterproof. This type of kayak is not intended for rolls (fig. 45).

The kayak frame must be made quite light if one intends to carry the boat to relatively inaccessible beaches. A builder wishing to use materials he has collected himself will need to exercise certain ingenuity. A wood-and-skin kayak is built like the fuselage of an old-fashioned aircraft with a stressed skin over a light frame. The frame consists of two long members (the gunwales) which range from 9 feet in the dumpy kayaks of Koriak Island to more than 22 feet in the kayaks of Baffin Island. In most cases these gunwales are made of spruce but the method of using interwoven willow or hazel, as in the British coracle, might well be tried by the experimentally inclined. In the Nunivak Island kayak the gunwales were attached in front to a carved stem piece made from a board on edge and at the stern to a similar stern piece. The ridge of the deck was made from a laminated double piece and the thwarts connecting ridge piece to gunwales were nine to a side and notched into the ridge piece and lashed. The ribs were commonly made of willow precurved and mortised to the gunwales. The longitudinal stringers were lashed to the outside of the ribs and carefully fitted to the stem and stern posts. The covering of these kayaks was made of bearded seal skin or sea lion skin. If these were not available walrus skin was used. The true Whole-earther, who wishes to stay on good terms with seals, sea lions and walruses, will probably prefer to use light canvas rendered waterproof by applications of boiled linseed oil and pitch. Or, if he does not disdain the products of modern technology he can make his kayak skin of fiber glass cloth stiffened with resin.

3. Other traditional watercraft

A remarkable and very ancient native boat is used by fishermen off the coast of Peru and illustrated in the *National Geographic* of March 1973 (p. 331). This craft, made of reeds and now called *el caballito,* was developed by the Chimu people and has been in use for 2,000 years. Making one of these would present an interesting challenge to any Whole-earther having access to a good supply of reeds. The most interesting feature of the boat is its upswept bow, which enables the Peruvian fisherman to take it through surf that would be difficult to negotiate in a kayak. This same upswept bow is, of course, a great disadvantage if one has to paddle against the wind as it considerably increases wind resistance (fig. 45).

In the same issue of the *National Geographic* will be found a photograph of the umiak, another Eskimo craft of great antiquity. The umiak, unlike the kayak, was an open vessel, might be as long as 60 feet and commonly was paddled by a crew of eight. Its most remarkable feature is the extreme lightness of the frame, consisting of a gunwale, two chines on each side and a central keelson held together by as few as 10 ribs to a side. The skin cover, of walrus hide, was stretched over this frame, passed over the gunwales and was lashed to the chines by means of ropes. This Eskimo boat resembles the Irish curragh, which is still in use in the Aran Isles and was undoubtedly originally a skin boat, cattle hide being used in the curragh whereas walrus hide was used in the umiak.

It would seem remiss to conclude this section on native watercraft without some mention of the birch bark canoe. These boats were the main means of transportation of the Micmac, Malecite, Algonquin, and Ojibway Indians. They were adopted by the fur traders and built like cargo boats, attaining lengths of 36 feet. The methods used for making these boats are entirely different from those used in building kayaks and are described at length in *The Bark Canoes and Skin Boats of North America*. Quite special skills are required both to gather the materials and to put them together. The art is by no means a lost one. Henri Villaincourt of Mill St., Box 199, Greenville, N.H. 03048, still makes these boats and his canoes (depicted in *The Last Whole Earth Catalog*, p. 283) have all the grace and beauty of the originals. The techniques of building these boats are too complex to be described here. Any Whole-earther wishing to master this craft might do well to serve an apprenticeship with Mr. Villaincourt. He says it takes a month's work to build one canoe and I can well believe it.

BODGING AND BOWLING

"Here and there were trim villages of timber and thatch, flint and brick, while above were the chalk pastures with grazing

298 CRAFTS

flocks of sheep. Crowning the hills were the glades of beech, shimmering green in the sunlight. Quite suddenly there was a gap in the trees and there in a clearing was the bodger's simple hut, surrounded by felled logs and hedgehog-like groups of drying chair legs and stretchers" (J. G. Jenkins, *The Traditional Country Craftsmen*).

All this has vanished. The last of the chair bodgers has died and the beech woods are silent. However, any true Whole-earther who wishes to make his own furniture using a bodger's lathe to carve chair and table legs can easily do so and get much pleasure in the process.

The bodger's lathe, or pole lathe, is a very simple piece of equipment of great antiquity. It consists of a long sapling (ash or larch) fixed at one end, free at the other. To the free end is attached a cord and a leather strap connected to a foot treadle below. A solid wooden frame with metal spikes at each end serves to hold the work in place (fig. 46). The strap is passed

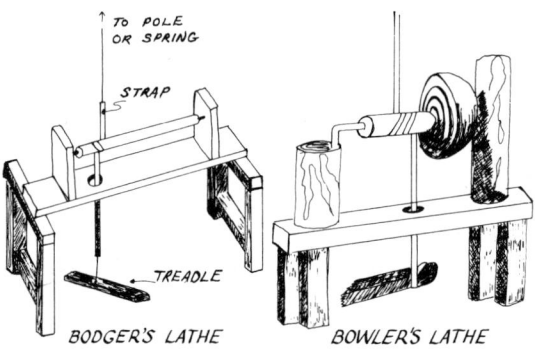

FIGURE 46

around the wood and down to the treadle in such a way that, when the treadle is depressed, the wood rotates *toward* the craftsman.

In working with a pole lathe the power stroke is the downstroke. The craftsman presses the treadle down and at the same time presses his chisel against the wood. He then releases the treadle, which is drawn up by the elasticity of the pole. The pole is a rather clumsy device and can be replaced by a spring or a weight slung over a pulley.

Admittedly the pole lathe is not very efficient because half the craftsman's time is wasted during the upstroke. It is, however, a satisfying rhythmic activity and the pole lathe costs practically nothing. The traditional English chair bodger always used beech wood for his chair legs. He selected straight-grained trunks, felled the trees, cut them into 18-inch segments with a cross-cut saw, split the segments into billets and shaped the billets roughly with a hand axe. Before being placed on the lathe the billets are shaved with a drawknife. Legs for traditional Windsor chairs were made with bobbin decorations cut with V-shaped chisels. The beech wood was worked while still green and the finished legs were stacked to dry for several weeks before being transported to the chair factories. A skilled bodger could turn out a chair leg in 2 minutes.

The American Whole-earther who does not have access to the Chiltern beech woods can do well with oak, hard maple or ash. The traditional Windsor chair had legs and stretchers of beech, a seat of elm and bow-shaped back of ash.

Very similar to the bodger's lathe was the bowler's lathe. In England there were bowl makers so expert that, using this crude equipment, they could turn as many as four bowls at a time from a single segment of elm. To do this they used gouges curved in such a way that they could remove a narrow segment of wood having a curved cross section. Elm was the preferred wood because it is almost unsplittable. The elm blocks were cut so that the grain ran across the opening of the bowl. Elm has to be carefully seasoned and needs about 5 years to dry before it can be worked. Here in California we find that the bay tree (pepperwood) is a good material for bowls if one can only find a piece that has not been ruined by wood worms. It is easy to carve, has an interesting grain and takes a nice polish.

PIT SAWING

Few Whole-earthers have used or even seen a pit saw. Nonetheless these instruments are available (The Loxley Saw, Sanderson Brothers, Newbould Ltd., Sheffield, England) and the art of using them is well worth learning. A pit saw is a very large rip saw designed for use by two men. The teeth have the form of chisels and must be sharpened as such. The saw is generally 7 feet long and 10 inches wide at the top, tapering to 3 inches at the bottom. The handle at the top is a simple T which fits permanently in the metal holder. The handle at the bottom is removable, as the saw must be removed from the cut to permit the log to be moved forward.

A log which is to be used for pit sawing should be allowed to season for a year. It should then be placed on a trimming frame and held in position with iron dogs while a flat surface is hewn on one side. Rotate the trunk and hew a second flat surface parallel to the first. The first surface provides a plane on which the log can rest, the second a means of marking the line along which the saw must be guided. These lines must be carefully spaced and allowance made for the width of the saw cut. They can be made by snapping a chalk line, using blue chalk if the wood is light, white chalk if it is dark. If the full width of the trunk is not required it is best to make a center cut first and then cut the two halves, resting the tree trunk on the sawn surfaces.

Although a pit is a traditional part of pit sawing it is not essential. If a fairly steep slope is available a stout frame can be built, bedded in the slope at one end. The tree trunk can be pushed out onto the support and secured by iron dogs to the frame. This arrangement eliminates two bad features of the pit: the bottom man hardly gets enough light to see what he is doing, and the pit tends to fill with water when it rains. A pit, however, is very convenient when really heavy trunks have to be sawn. Such massive entities can be manipulated into place with crowbars and no heavy lifting equipment is needed. The pit should be deep enough to allow a man to stand upright without hitting his head on the tree trunk he is sawing. It was traditionally about 15 feet in length, had two logs lying along the full length on top and shorter pieces holding them in place at the ends. These were called strakes and sills.

Success with a pit saw depends on the first few minutes of work. Only at the beginning can the direction of the cut be changed. Once the saw is fully into the wood corrections become very difficult. The work involves close harmony between the top man, who must guide the saw, and the bottom man, who, working in a steady stream of sawdust, must follow the directions by feel and avoid imparting any twist to the big blade. For two people working well together the exercise can be a great pleasure. But when things go wrong, when cuts get out of line or out of perpendicular, much friction can be generated both physical and emotional.

After a tree trunk has been sawn into planks it must be further seasoned (1 year for every inch of thickness). For this purpose the planks should be separated by a series of slats wide enough to allow air to circulate between them.

BLACKSMITHING

The craft of working in iron is one of the oldest and most honored. In ancient times the blacksmith was commonly thought to possess magical powers. In medieval Wales the smith took a place of honor along with the priest and poet in the Prince's Court.

Today, of course, all this has changed. The infinitely varied parts required by today's machines are formed from countless special alloys by electronically controlled machine tools receiving their instructions from coded punch cards. The art of the smith is confined almost entirely to the creation of wrought-iron work and the shoeing of horses.

For the true Whole-earther working in iron may be a deeply satisfying experience, even though the economic returns from such activity may be small. A blacksmith's shop naturally centers on the forge or hearth, which can be large or small depending on the size of work to be done (fig. 47). The vital part of the hearth is, of course, the bellows, by means of which a blast of air is forced into the heap of anthracite that supplies the heat needed to soften the iron. These bellows were sometimes as much as 7 feet long and had a handle so arranged that the smith could operate it with his left hand while manipulating the iron with his right. Of course, one who is not reluctant to rely on electrical power will settle for an electrically driven blower.

During the centuries the blacksmith has evolved quite a com-

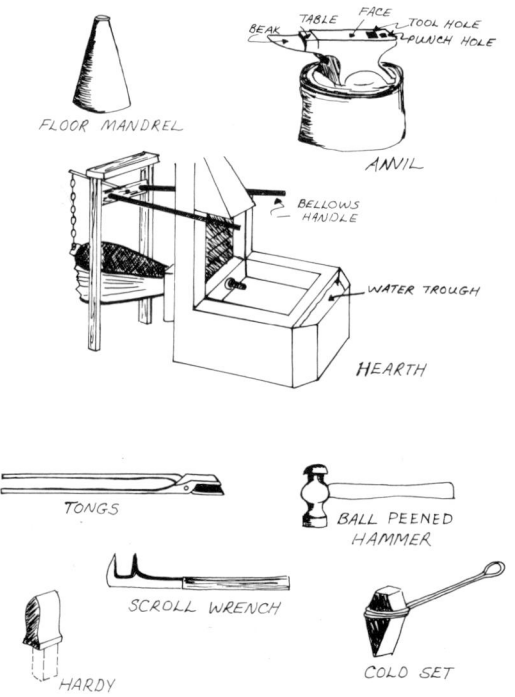

BLACKSMITH'S TOOLS

FIGURE 47

plex array of tools with which to work iron. Perhaps the most vital is the anvil. It has been said, "Working on a bad anvil is like jumping on a heap of sand, whereas working on a good anvil set on a proper foundation is like jumping on a springboard." The "London pattern" anvil has a hardened steel top but the table and beak are of wrought iron. The blacksmith, when cutting a piece of iron, moves the work to the table before striking the final blow to avoid damaging the chisel.

In addition to the anvil a blacksmith needs a heavy bending vice and a smaller vice. Tongs of various shapes and sizes are necessary for holding the iron. Cold chisels are used to cut the iron when cold, and cold sets, with handles of twisted wire, are used with a sledge hammer version to make deeper cuts. Hot chisels and hot sets are used for cutting the hot iron. A hardy, which is a heavy chisel that fits into the square tool hole of the anvil, may be used as a hot chisel. Various punches are used for making holes in the hot metal and these holes may be enlarged by means of tapered steel drifts. In addition to a sledge hammer for heavy work the blacksmith uses a number of special lighter hammers, the form of which depends on the work to be done.

The chief characteristic of wrought iron is that it has a grain. Its fibers lie in one direction and its strength lies along the fibers. Wrought iron, therefore, must always be forge-worked. It cannot be turned on a lathe because this would destroy the fibrous structure. In this respect it differs from mild steel, which is non-fibrous and equally strong in any direction. The main operations in working wrought iron are as follows:

(1) Upsetting. This involves increasing the thickness of a bar by striking the white-hot iron vertically, end-on to the anvil. (2) Drawing down. The thickness of the iron may be decreased and its shape may be altered. This is done by the use of fullers, flatters and swages, all of which are used in pairs, the lower one being placed in the tool hole of the anvil, the upper one being struck with the hammer. (3) Bending and twisting. These operations can be done when the iron is at cherry-red heat. A curve may be produced by hammering the iron on the beak of the

anvil. Scrolls, which figure so largely in wrought-iron work, are produced by fixing one end of the work in a vice and twisting the other by means of a scroll iron. (4) Welding. By this operation two pieces of iron are joined together by hammering while they are at white plastic heat. (5) Riveting. This is a second method of joining two pieces of iron. Holes are cut in each of the pieces to be joined and a white-hot rivet is introduced and hammered flat at each end.

INDEX

Page numbers in italics denote illustrations

A

Abalone, cultivation of, 64, 65, 66
Acid, acetic, 99, 202; amino, 186, 196; corrosive, 239; hydrochloric, 186; lactic, 101, 190; sulfuric, 119
Acne, 226
Adobe brick, building with, 148
Adrenalin, 188, 193
Air, 95; circulation of, 140, 142, 150; contamination through, 205; polluted, 201, 202. *See also* Power, wind
Airway, blockage of, 233
Alcohol, abuse of, 203, 221; ethyl, 94; as poison, 202–03
Algae, 63, 64, 181
Alkalis, corrosive, 239
Allen, Edward (*Stone Shelters*), 148
Americans, food consumption by, 199
Anemia, 224–25
Angina pectoris, 215
Angling, 76
Antibodies, 186–87
Antihistamines, 213
Apoplexy (stroke), 218
Appendicitis, 212
Appestat, 200, 201
Aquacoms, 177–78, 180–81, 259

305

INDEX

Aquaculture, 55–67
Aquifers, 170, 171, 172, 173
Arrowhead (wapatoo), harvesting of, 79
Arthritis, rheumatoid, 223
Asphyxia, 245
Asthma, 215
Atherosclerosis, 200
Atlatl (spear thrower), 70–71
Axes, stone, 69–70

B

Bacteria, 87, 91, 101, 111, 205, 206, 214, 225, 270
Barbecuing, 111
Bark Canoes, The (Adney and Chapelle), 295, 297
Barley, 34–35; malting, 98
Basement, *see* Foundations
Basket making, 277, 278, 279–81
Baskets, coiled, 281; hamper, 278–79; osier, 278–80; split oak, 277–79
Bate, Harold, 271, 272, 274
Beadwork, 289–90, 291, 292
Beans, 27–29, 109; baked, recipe for, 109–10; dried green, 86
Beds, raised garden, 13–14
Beef, 199; pickled, recipe for, 89–90; standard cuts of, 92
Beer, 35, 202, 203; brewing of, 98–99
Beetle, flea, 41, 46; Mexican bean, 29
Beriberi, 36, 198
Berries, 48, 78
Biscuit (pottery), 283
Bites, poisonous, 241
Blackout, 219
Blacksmithing, 301–04; tools of, 302

Bladder, urinary, infection of, 193, 223
Bleaching, of yarn, 128
Bleeding, 234
Blight (late), 37, 38
Blisters, 237
Blood, 186, 187, 188, 193; clots of, 217; disorders of, 224–26; supply to muscles, 190
Blood pressure, *see* Hypertension
Boards, batter, 143, 144
Boats, building, 178–79; reed, 76, 292–93, 294, 296; skin, 293
Bodgers, of chairs, 298, 299
Bones, 91, 191–92; broken, 236; disorders of, 223–24; marrow of, 192
Bordeaux mixture, 38, 48
Borsook, Dr. Henry, and MPF, 93
Bottling, 86–87; of wine, 97
Botulism, 87, 206
Bow, how to make, 70
Brace Research Institute, 257, 259
Brain, 188, 189, 192; of animals, as food, 110
Braising, of meat, 111
Brassicas, 40–41
Bravery's Super Stout, recipe for, 98–99
Bread, baking, 115–16; recipe for, 100; steamed, 101; substitutes for, 109
Bread making, 31, 32, 99–101
Breakdown, of bodily health, 196–208
Breast, cancer of, 230
Brewing, 98–99
Bronchitis, 213–14
Brose (oatmeal), 34
Building, materials for, 144–49
Bulbs, 42–43
Burns, 237–38

Butchering, 91
Butter making, 105

C

Cabin, 139, 141; log, 149–56; stockade, 153
Calcium, 197–98
Calico, unbleached, 293
Calories, 38, 197, 199–201
Campden tablets (sulfite), 94
Cancer, 202, 214–15, 230–31
Canning, 86–89
Canoe, 76; bark, 293, 297; dugout, 293
Car, run on methane gas, 274–75
Caretakers, group, of land, 2–4
Casing, for wells, 172
Casting, slip, 284
Catfish, 62–63
Caves, as homes, 144, 145
Cells, photovoltaic, 252–53
Cerebral palsy, 219–20
Cervix, cancer of, 230
Cheese, aging, 103–04; cheddar-type, recipe for, 102–03; molds for, 104; soy, 104
Cheese making, 96, 101–05
Chemicals, in body, 184; in eye, 240; as fertilizers, 21, 22
Childbirth, 242, 243, 244–45
Children, and bronchitis, 213; and choking, 233; and convulsions, 218–19; and heart disease, 216–17; immunization schedule for, 208; and impetigo, 226; intestinal worms in, 212; and mild dysentery, 209–10; and nephritis, 221–22; and rheumatic fever, 217
Chili con carne, recipe for, 112

Chimney, 166, 168; waterproofing, 167
Choking, 233
Cholera, 210
Circulation, disorders of, 215–17; systems of, 187–88
Cirrhosis, 221
Cities, floating on water, 181
Clay, 281, 282, 283; firing, 288–89; working of, 283–87
Climate, and bronchitis, 213; and house designs, 139, 141–42
Cloth, 122–30
Clotting factor, in blood, 225
Clover, sweet, 12, 29, 36, 52
Colds, common, 212–13
Cold sores, 227
Colitis, ulcerative, 211
Coma, 218, 219, 245
Compost, 9, 10–11, 40, 46, 61, 175, 176
Compressor, for gas, 274
Constipation, 211
Construction, with earth, 145–48; methods of, 146; post-and-beam, 158, 159
Contamination, transmittal agents of, 205–07
Convulsions, 218–19
Cookers, fireless, 116–17; solar, 108, 114, 115. See also Pressure cookers
Cooking, 105–17; meat, 111–12; fish, 112–13; solar, 112, 114
Coracles, 293, 294
Corn, 33–34
Cramps, from heat, 238
Crops, major, 26–42; minor, 42–46; rotation of, 30, 53
Cucurbits, 39–40
Culture, hydroponic (soilless), 24, 25; Laver, 64; of ocean fish, 67

Curds, 101, 103; soybean, 105
Curragh, 297
Cuts, standard meat, 92
Cyanide, 240
Cystitis, 223

D

Dams, 266, 267
Davis, Adelle (*Let's Cook It Right*), 108, 110, 111
Death, 245–47
Deer, hunting of, 71
Defenses, of the body, 186
Depression, 232–33
Derris root, 59–60
Desert, survival in, 78, 82, 259
Designs, for aquacoms, 180; for houses, 138–42, 165
Diamond, Steve (*What the Trees Said*), 6–7
Diarrhea, summer, 210–11
Dickey, Esther (*Passport to Survival*), 92–93
Diet, 200, 201
Diet for a Small Planet (Lappé), 199–200
Digester, sealed, 271, 272, 273
Digestion, process of, 184–86
Disease, carriers of, 206, 207; immunization against, 207–09; infectious, 227; prevention of, 204–07; venereal, 228–30
Dislocations, of joints, 235–36
Distillation, of sea water, 25, 82; solar, 180, 181, 256. See also Stills, solar
Domes, 142
Donaldson, Dr. Lauren P., and fish culture, 67

Dormers, 163, *164*
Dougherty, John W. (*Pottery Made Easy*), 288
Dowsing, 170
Drainage, 143; of waste, 174, **177**
Dreams, 189
Drill, auger, 172, 173; bow, 83; seed, 31
Drowning, 234
Drugs, depressant, 239; as poison, 203–04
Drupes, 48
Dry foods, natural, 91–93
Dumplings, recipe for, 101
Dyeing, 128–29
Dysentery, 209–10

E

Earth, building with, 145–48
Earthenware, 282
Eating, proper way of, 108
Eczema, 228
Edible Wild Plants (Medsger), 81
Eggs, 53; fish, 62
El caballito, *294*, 296
Embolism, arterial, 217
Emergency Medical Guide (Henderson), 237
Emphysema, 202, 213–14
Endocrine system, 192–93
Enema, 212
Energy, alternate sources of, 248–75
Energy crisis, 249–50
Enzymes, 112, 184, 186, 198, 202
Epilepsy, 219
Erosion, 16, 54; prevention of, 61
Eskimos, 200, 295, 297
Espalier, method of, *38*, 47
Euthanasia, 246–47

F

Faulkner, Edward H. (*Plowman's Folly*), 11
Female sex organs, *194*, 195–96
Fermentation, 32, 41, 93–105, 271
Ferrocement, 179
Fertility, of fish ponds, 57; of soil, 8–9, 54
Fertilizers, 21–23, 39; in fish ponds, 59, 62, 63
Fever, rheumatic, 217
Fiber, from plants, 74–75, 122; of wool, 128
Fiber glass, 179
Filtration, of wine, 98
Fire, 82–84, 105
Fireblight, 47–48
Fire bundle, 83–84
Fire pit, 115, 116
Fireplaces, 166–67
First aid, 233–41. See also Individual accidents
Fish, cooking of, 112–13; cultivation of, 55–67; drying, *85*; feeding, 60; in rice paddies, 36; smoking, 84, *85*
Fish Culture (Hickling), 56–57
Fish-pops, recipe for, 113
Flax, 122, 124; preparation of, *123*
Floors, for frame house, 157–60; for log cabin, 152
Flour, 32, 99
Flowers, of legumes, 27
Folding, of livestock, 52
Food, animals as, 51–53; contamination of, 206–07; gathering of, 68–84; growing of, 2–67; poisoning, 209; spoilage of, 86–87; storing and processing, 84–117; texture of, 108; waste of, 26
Foraging, on shoreline, 73–74
Foundations, laying of, 143, *144*; for log cabins, 149–50, 151; for frame house, 157; of stone, 149
Foxfire Book, The (Wigginton), 152, 154, 156, 276
Fractures, of bones, 236
Frame house, building of, 156–65
Framing, on a house, *158*
Freezing, 90–91
Fritatas, recipe for, 113
Frostbite, 238
Fruits, 46–51; drying, 86
Fuel, 165; minimum of, 107–08. See also Power; Energy
Fungus, 241; of skin, 227
Furniture, wickerwork, 279

G

Gall bladder, 220–21; inflammation of, 211
Game, 112
Gangrene, 217, 238
Garden, 3, 7, 23; vegetable, 54, 55
Gardening, organic, 22
Gas, poisonous, 240; methane, 174, 181, 270–75
Gastrointestinal tract, disturbances of, 209–12
Generators, 263, 264, 265, 270, 271, 272; solar, 252
Gentle Giants, The (Jepson), 17
Gerwick, Ben C., Jr., and aquacoms, 181
Gibbons, Euell, 73, 74
Glazing, on pottery, 288
God, land given to, 4–6

Gonorrhea, 229
Grains, 98
Grapes, 48–49
Grasses, 29–37
Gravity, specific, 96–97; and water, 168–69
Gray, Asa (botanist), 80
Greenhouse, 23–24, 39, 44, 45, 259
Greenland, and kayaks, 295
Greenware, 283, 388
Growing pit, 23, 24

H

Hawaiian Islands, 180
Hay fever, 213, 215
Heart, 186, 187; diseases of, 202, 216–17; as muscle, 190; problems of, 215–17; strain on, 214
Hearth (forge), 302
Heat, 106, 190; generation of, 25; solar, 82; treatment from, 107. See also Generators; Power, solar
Heaters, solar, 253, 254
Heating, of caves, 145; of compost, 11; for houses, 165–68; problems of, 165
Hemoglobin, 187, 224, 225
Hemophilia, 225
Henderson, Dr. John (*Emergency Medical Guide*), 237
Hepatitis, 221
Herbs, 44–45
Hickling, C. F. (*Fish Culture*), 56–57
Hiroshima Bay, oysters in, 65
Hormones, 187, 192, 193, 195, 196
Horses, 14, 16–20

Hot frame, 23, 24
Houseboats, 177, 178–79
Houses, designs for, 138–42, 165; sites for, 142–43; solar heated, 140, 255, 256; on water, 177–81
Housing, 130–81
Hunting, 69–71
Hush puppies (Florida), recipe for, 100
Hydraulic ram, 169
Hydroponics, 24, 25
Hypertension, 216

I

Illumination, in caves, 145
Immunization, 207–09
Impetigo, 226
Impulses, social, 138
Indians, 79, 289, 297. See also individual tribes
Infants, and diarrhea, 210–11; and impetigo, 226; and rickets, 224
Influenza, 212–13; gastric, 209
Insecticides, 41, 59
Insects, contamination by, 207
Insulation, from cattail, 81; of houses, 168; sheathing, 160
Intestines, 186, 220–21
Invertebrates, marine, 65–66
Irritants, poisonous, 240

J

Jam, making of, 90
Jaundice, 220
Jellies, making of, 90

INDEX

Jepson, Stanley M. (*The Gentle Giants*), 17
Jerky (dried beef), 84, 85

K

Kayaks, 75–76, 293, 294, 295–96
Kern, Ken (*The Owner-Built Home*), 131, 139, 156, 165, 169, 174, 176
Kick wheel (potter's), 286, 287
Kidneys, 193; diseases of, 221–23
Kilns, 287, 288–89
Knots, for rugs, 130
Kwashiorkor, 41, 197

L

Land, buying, 6–7; given to God, 4–6; rental of, 6; size of, 7
Lappé, Frances Moore (*Diet for a Small Planet*), 199–200
Last Whole Earth Catalog, The, 64, 70, 83, 99, 129, 148, 228, 242, 297
Lathes, pole, 298, 299
Laubin, Reginald and Gladys (*The Indian Tipi*), 133–34
Layout, of crops, 53, *54,* 55–56
Leaching, by cooking, 106
Leaks, sealing in ponds, 58–59
Legs, and varicose veins, 217
Legumes, 21–22, 26–29, 36; wild, 78, 80
Lesch, Alma (*Vegetable Dyeing*), 128
Let's Cook It Right (Davis), 108, 110
Leukemia, 225–26
Lice, 227
Ligaments, 235

Lime, hydrated, 58, 273; quick, 60; slaked, 120, 176
Linen, 122
Liver, 110, 186, 202, 203; disorders of, 220–21
Livestock, 10, 18–19, 36, 51, 52–53
Living the Good Life (Nearing), 51, 149
"Living Will," 246–47
Loom, 129–30, 289, *291;* simple frame, *127*
Lummi Indians, and oysters, 66
Lungs, 186, 187, 202; cancer of, 214–15

M

Male sex organs, 193, *194*
Malnutrition, 34, 36, 197
Manure, 10, 22, 23, 40, 47, 52, 59, 61, 108, 205–06, 271, 272
Marinating, 112
Maturation, of wine, 97
Measles, 208; German, 220
Meat, drying, 84–85; tenderized, 112
Medsger, Oliver P. (*Edible Wild Plants*), 81
Meningitis, 218
Menstruation, cycle of, *195, 196*
Mental illness, 231–33
Methane gas, 174, 181, 270–75
Milk, 101, 105, 198, 199; aquatic, 180; powdered skim, 93; and protein, 52; soy, 104
Mill, hand, 32; roller, 34
Minerals, in diet, 197–98
Moccasins, 121–22
Molds, for cheese, 104
Mongolism, 220

MPF (Multi-Purpose Food), 93
Mulch, surface, 11–12, 14
Mules, 14, 20–21
Mumps, 208
Muscles, 189–91; back, 235; contraction of, 185; coordination of, 203
Mushrooms, 46; poisonous, 241
Mussels, 74
Must, of wine, 94, 95

N

Navahos, 130
Nearing, Helen and Scott (*Living the Good Life*), 51, 149
Nephritis, 221–22
Nervous system, 188–89, 190; damage to, 203; disorders of, 217–20
Nets, 72, 74–75
Nitrogen, 12, 27, 40; shortage of, 33
Notching, 154
NPK (Nitrogen-Phosphorus-Potassium), 21, 22
Nunivak Island, and kayaks, 295, 296
Nuts, 48, 50; wild edible, 76–77, 78

O

Oak, bark for tanning, 120; splits of, 277
Obesity, and blood pressure, 216
Olsen, Larry Dean, 68, 69, 70, 71, 82
Organisms, pathogenic, 10, 173, 175, 177, 207
Osteoarthritis, 223–24

Outdoor Survival Skills (Olsen), 68
Outward Bound School, 69
Oven, brick, 100; Dutch, 116; warm-air, 86
Owner-Built Home, The (Kern), 131, 139, 165
Oxygen, in water, 59
Oysters, cultivation of, 64, 65, 66

P

Paiutes, 74, 77, 120
Pancreas, 193, 221
Paralysis, 218
Paranoia, 218
Parasites, 204–05, 206
Parkinsonism, 219
Pasture, 36, 52, 53
Pemmican, 85, 106
Petroleum, distillates of, 239
pH, of water, 57–58
Phytoplankton, 64, 65
Pickles, dill, recipe for, 90
Pickling, 89–90; of skins, 119
Piers, 150, 157, 158
Pit sawing, 300–307
Pituitary gland, 192, 195, 196
Pit vipers, 241
Placenta, 244
Plants, poisonous, 241; of sea, 64–65; wild edibles, 76, 77, 78–82
Plow, moldboard, 11, 13, 14, 15; horse, 16
Plowing, 9, 15–16
Plumbing, in caves, 145. *See also* Waste, disposal of
Pneumonia, 214
Poison, 87; on arrows, 71; in chokecherry, 79; fish, 59–60; intake of, 201–04, 206; ivy, 228

Poisoning, 239–41; food, 206
Pollens, 213
Pollination, cross, 39; by wind, 33
Pollution, industrial, 67; air, 201, 202
Pomes, 46, 47–48
Pond, construction of, 58; for fish supply, 57–60
Population, zero growth of, 138
Porcelain, 282
Pork, standard cuts of, 92
Potatoes, 37–38, 39; sweet, 41–42
Pottery, 281–89; slab, 284
Power, 90, 249; electric, 32, 250, 263; methane, 270–75; solar, 113–14, 251–59; water, 265–70; wind, 259–65
Press Democrat (Santa Rosa), 3
Pressure cooker, 87, 88, 89, 111, 116
Pressure points, 234
Privacy, 138–39
Privy, 176, 177, 205
Processing, times of, 88
Protein, 35, 36, 105, 107, 112, 197, 199–201; animal, 51; conversion of, 51–52; deficiency of, 109; in vegetables, 28, 29, 34, 37
Pruning, 47
Psychedelics, natural, 204
Psychopath, traits of, 232
Psychosis, manic-depressive, 232
Puberty, 192
Purgatives, 211–12
Pyelitis, 222
Pyrethrum, 29, 41, 46

Q

Quicksand, 172

R

Rabbits, 52; skins of, 120–21
Rabies, 208
Raceways, for water, 60–61, 62
Racking, 96–97
Rafters, 163
Rafts, 76, 292
Rammed earth, for houses, 147–48
Reactor, breeder, 250–51
Recycling, 65, 180
Reflectors, solar, 114
Reproduction, organs of, 193–96
Reservoir, 169
Respiration, failure of, 233–34; process of, 186–87
Respiratory system, disturbances of, 212–15
Resuscitation, 233, 234
Rice, 35–36, 198
Rickets, 224
Ringworm, 227
Riveting, 304
Roads, proximity to, 143
Rock, 8, 144–145, 172
Roofs, beam-and-plank, 161; Gambrel, 163; framing, 162, 164; shed, 161; thatched, 163–64
Rooms, arrangement of, 141
Roundworms, 212
Rugs, knotted, 130
Rustrum, Calvin (*The Wilderness Cabin*), 141, 153

S

Saliva, 184, 185
Salt, 25, 89, 90, 198
Sauerkraut, making of, 41

Savonius rotor, building of, 260, *261*, 262
Saw, chain, 149, 152, 153, 156; pit, 149
Sawmill, 156
Scabies, 226–27
Schizophrenia, 231
Sciatica, 220
Scurvy, 198
Seaweed, 64, 74
Sego, 79–80
Septic tank, 174, 175, 177
Sewage, 22–23, 59, 61, 63, 65, 181, 205
Sgraffito, 287–88
Shakes, laying, 161, *162*
Sheep, 124–25
Shingles, 154, 156, 161, 162, 227–28; making and laying, *155*
Shock, 237; anaphylactic, 240
Singh, Dr. Ram Bux, 271
Sioux, 80, 121
Skeleton, 191–92
Skin, cancer of, 230; disorders of, 226–28; grafts of, 237
Skin Boats of North America (Adney and Chapelle), 295, 297
Skinning, 91
Skins, 118–22
Skull, 191
Sludge, in wine, 95
Slurry, 120, 273, 283
Smoking, of cigarettes, and cancer, 202, 214
Snakebite, 241
Snare, 72
Sod, for houses, 147
Soil, composition of, 8, 9; contamination of, 205; potting, 24; preparing, 7–25; texture of, 9–10; working of, 12–14
Solanums, 37–39

Sorgo syrup, 36–37
Sorter, of wool, 124
Sowing, broadcast, 30–31; corn, 33
Soybeans, 28, 93, 104, 109, 199
Spacer, for sowing, *56*
Spiders, black widow, 241
Spillway, 58, 267
Spinal column, 189, 191
Spindle, 126
Spinning, 122; wheel for, *127*
Sprains, 235
Spring house, 169
Stalking the Wild Asparagus (Gibbons), 73
Steam, harnessing, 265
Sterilization, of bottles, 95; of food, 106
Stewing, of meat, 111
Stills, solar, 181, 256–57, *258*, *259*
Stings, insect, 240
Stomach, 186; ulcers of, 202, 211
Stone Shelters (Allen), 148
Stoneware, 282
Storage, of heat, 114; of water, 169, 171, 253; of potatoes, 38
Stoves, 166, 168
Strains, 235
Strapping, supportive, 235
Stress, and ulcers, 211
Sugar, 90, 95
Sulfuring, 86
Sun, as heat source, 83, 86. See also Power, solar; Stills, solar
Sunstroke, 238–39
Syncope, 245
Syphilis, 229–30
Systems, cardiovascular, *185*, 186–87; digestive, 184, *185*, 186; endocrine, 192–93; muscular, 189–91; nervous,

188–89; reproductive, 193–96; respiratory, 186–87; skeletal, 191–92; urinary, 193. See also individual organs

T

Tanning, rawhide, 120; skins, 119
Tapeworms, 212
Telkes, Dr. Maria, 114
Temperature, control of, 111; of water, 57
Tendons, 190–91
Tension, and alcohol, 203; and blood pressure, 216
Tent, 131, 133, 138
Tetanus, 206
Throat, sore, 213
Thrombosis, coronary, 215–16
Thyroid gland, 193
Tipi, 131, 133–34, 135, 136–37, 138
Tobacco, as poison, 201–02
Toilet, 173, 174, 176
Tonsillitis, 213
Tools, soil-working, 14–16; hand, 153–54
Tortillas, recipe for, 100
Tourniquet, 235, 241
Toxins, see Poisons
Trapping, 71–73
Traps, deadfall, 71; eel, 75; fish, 75; for sun, 251; sunken, 72; examples of, 72
Trees, for building, 149, 151, 152, 156
Trenching, 9
Trichinosis, 111, 207
Tropins, 192
Trout, 61–62

Tuberculosis, 214
Tugboats, 178

U

Ulcers, 202, 211
Umbilical cord, 244
Umiak, 297
Urinary system, 193; disorders of, 221–23

V

Vaccines, 208
Vegetable Dyeing (Lesch), 128
Vegetables, 37–42, 110, 198, 199; drying, 86; dyes from, 128; leafy, 43–44, 45; luxury, 45–46; starchy, 37–38, 39, 107. See also Individual groupings
Veins, 234, varicose, 217, 242
Vinegar making, 89, 99
Virus, 29, 205, 208–09, 212, 214, 221, 227, 230. See also Organisms, pathogenic
Vitamins: A, 44, 113, 198; B, 112, 198; C, 39, 78, 93, 110, 198; D, 113, 198–99, 224; E, 199; in heavy doses, 231

W

Waffles, recipe for, 108–09
Walls, for frame house, 160
Wastes, animal, 270, disposal of, 173–77; radioactive, 250; units for, 175
Water, 25; balance in body, 192; contamination of, 205; depth

Water (cont.)
 of, 57, 58; flow of, 57, 62; living on top of, 177–81; pumping of, 260, 263; raising, 169–75; sources of, 171; supply of, 143, 168–73; to survive, 82; temperature of, 56, 59, 61
Watercraft, 292–93, 294, 295–97
Waterproofing, 148, 161, 293, 296
Water table, 170
Water wheels, 170, 265–67, 268, 269, 270
Weapons, primitive, 69–71
Weaving, 129–30; diagonal, 290–91, 292; loom, 289–90
Weeds, 8–9, 30
Welding, 304
Wells, 170, 171, 172
What the Trees Said (Diamond), 6–7
Wheat, 30–32, 34
Wilderness Cabin, The (Rustrum), 141, 153
Willows, reeds of, 75, 279, 280, 293
Wind machines, 260, 262–64
Wine making, 50, 93–95, 96, 97–98
Wool, 53; sorting of, 123, 124, 125
Worms, 61, 63; intestinal, 212
Wrought iron, working, 303–04

Y

Yarn, 126, 128
Yeast, 31, 32, 35, 94, 99
Yogurt, 101
Yurt, 131, 133, 137, 138

Z

Zoning regulations, 143

ROBERT S. DE ROPP

Dr. de Ropp is a biochemist, writer and guru. Formerly a visiting investigator at the Rockefeller Institute, he has carried out research in the fields of cancer, mental illness and drugs which affect behavior. His books, which have influenced people of all generations, especially the young, include *The Master Game, Drugs and the Mind,* and *The New Prometheans.* In his book *Church of the Earth,* which describes his experiences in setting up a rural commune, he looks over "the lattice of his life" and finds that he has launched "three children, two houses, six kayaks, seven books, forty-five scientific papers, a fertile garden, and fifteen fruit trees."